GENERAL CHEMISTRY 2 Laboratory

CHM 2046L

ELEVENTH EDITION

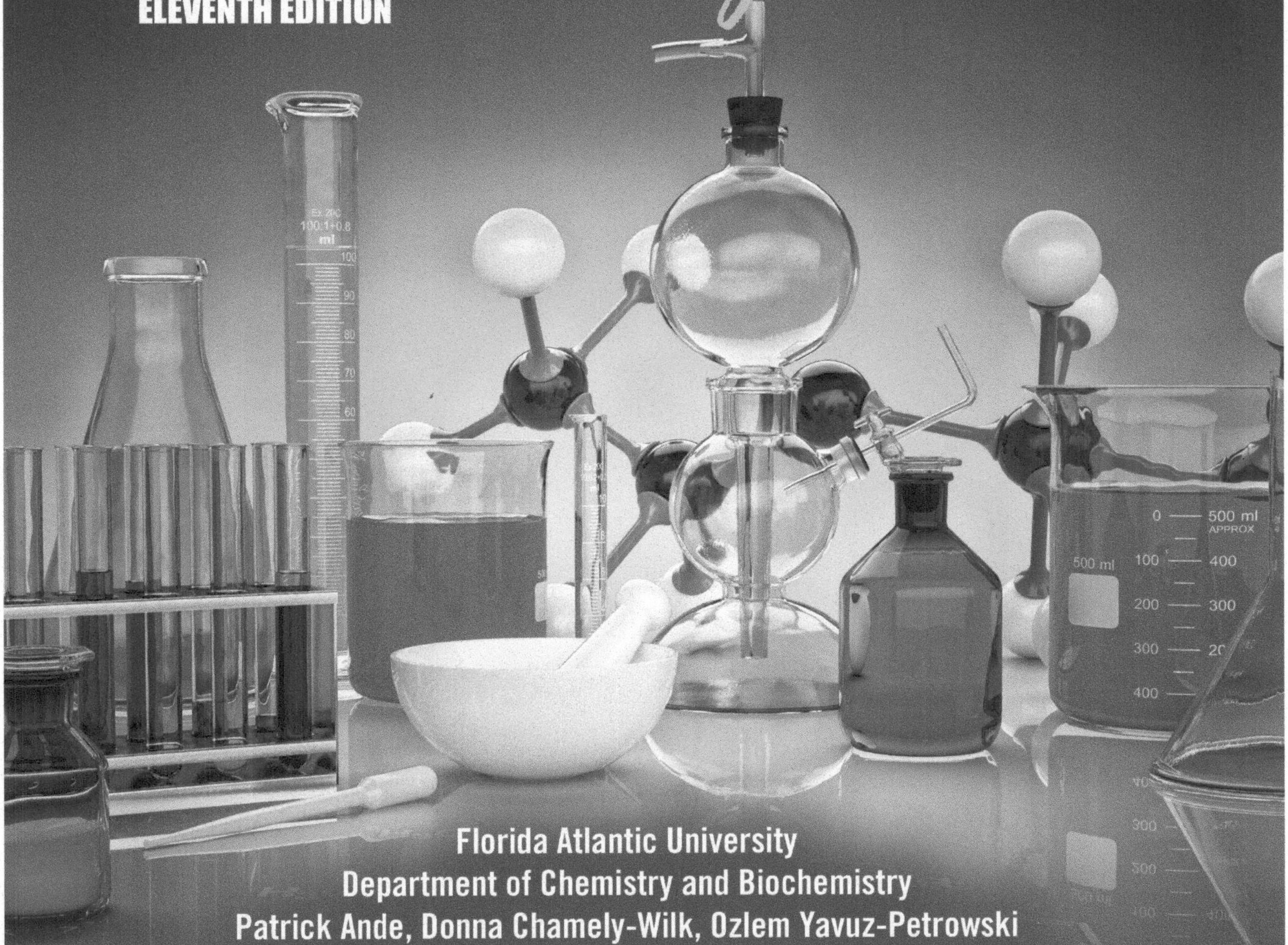

Florida Atlantic University
Department of Chemistry and Biochemistry
Patrick Ande, Donna Chamely-Wilk, Ozlem Yavuz-Petrowski

Kendall Hunt
publishing company

All images were redrawn by Kendall Hunt Publishing.
Cover image © Shutterstock, Inc.

Kendall Hunt
publishing company

www.kendallhunt.com
Send all inquiries to:
4050 Westmark Drive
Dubuque, IA 52004-1840

ISBN 978-1-7924-2951-4

Published in the United States of America

Contents

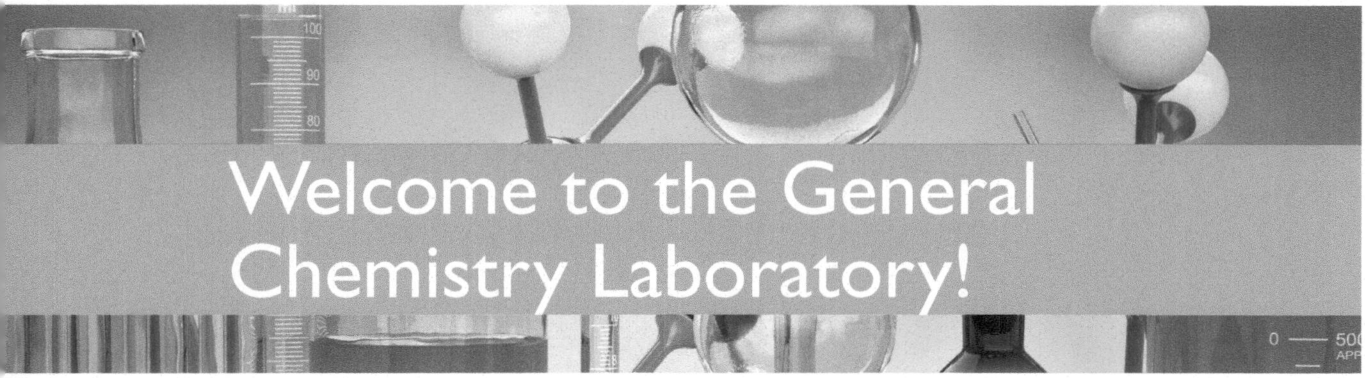

Welcome to the General Chemistry Laboratory!

Welcome to General Chemistry II Laboratory. Chemistry has the potential to be one of the most exciting courses you take in college. It is also a course that requires additional work and effort on your part. The laboratory is designed to support and illustrate chemical concepts studied in the lecture portion of the course, and give hands-on experience working with chemicals and laboratory equipment. The course objectives are as follows:

Course Objectives

- To perform experiments illustrating key concepts in chemistry
- To master basic laboratory techniques used in General Chemistry laboratories
- To get hands-on experience in chemistry
- To reinforce material presented in the CHM 2046 lecture course
- To begin learning how to write laboratory reports

Course Website

The Canvas site for this course, where students can obtain the course syllabus, grades, and specific information related to their section, as well as announcements regarding individual laboratory experiments, can be reached at http://canvas.fau.edu. Your username is the same as your FAUNet ID and your password is the same as your PIN number.

Lab TA's

At the first meeting, you will meet with your teaching assistant (TA), who will supervise your lab work and grade your lab reports. Office hours for your section will be announced at your first lab; however, we encourage you to attend **any** Chem II TA's office hours. The schedule and location of these office hours will be posted on Canvas and in all of the labs. Please note that the TA is not your private tutor; they are there to facilitate the learning process.

We wish you a safe, successful, and enjoyable course!

Contact Information

Photo © 2014 Scott Silversten.

Laboratory Director

Patrick J. Ande
Office Location: PS 214
Office Phone No.: 297-2612
Office Hours: T, W, Th 10 a.m–12:30 p.m
E-mail Address: pande@fau.edu

Department of Chemistry and Biochemistry

Office Location: PS 110
Office Phone No.: (561) 297-3390
Office Fax No.: (561) 297-2759

Safety Rules

Dress Code

Certain attire is required in the General Chemistry Laboratories. **If you are not dressed according to the dress code, you will not be allowed to stay in the lab and you will receive a zero for that lab.** Students attending the lab should be dressed according to these rules:

- Long hair must be tied back.
- Wear safety goggles and/or glasses, approved by the department, at all times while in the lab and while

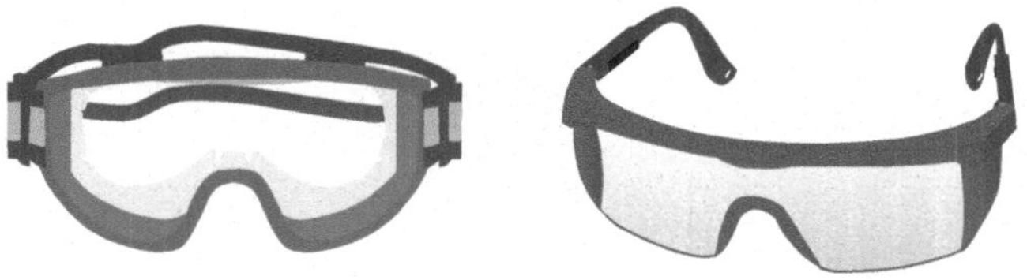

  handling any chemicals or equipment.
- Clothing must cover and protect the shoulders, upper arms, and entire upper body. Stomach and back should also be covered.
- Long pants or very long skirts are required. Legs have to be covered **down to the ankles**. Socks are strongly recommended.
- Shoes must be worn and have to cover **the entire feet**. **NO** sandals, open-toed, open-front, or open-back shoes are allowed.
- No students may wear shorts or a short skirt or have any portion of his/her midriff exposed, if you desire to wear a lab coat. Your TA can advise you where an appropriate lab coat can be purchased.
- Use of apron and/or gloves are not required, but are encouraged. These are provided by the lab if desired by the student.

Other Regulations

- Students are only allowed to enter the laboratory if a TA or the lab director is present.
- **Do not sit on the workbench** as it might have residues of previous chemical spills that can harm you. Keep your workstation neat and clean to avoid harming others.
- No food or drink in the lab to avoid possible contamination.
- Keep your hands away from your face while working, and wash them as often as possible.
- Be well prepared before you come to the lab.
- Know what to do in case of an emergency.

Accidents and the way to handle them: ALWAYS NOTIFY YOUR INSTRUCTOR!

- Emergency equipment:
  - Phone: PS 214, lab director's office, PS 208 and PS 207, prep room
  - Fire extinguishers: two in each lab, beside the doors
  - Emergency shower and eyewash station: in the SW (210 and 204), NW (209), or SE (206) corner of the lab
  - First aid kit: located on the top shelf of the middle bench, for small cuts and burns
  - Health Services Office on campus

- Spill kit: located in each prep room (PS 207, 208)
- MSDS sheets: located in each prep room (PS 207, 208)

- Broken glass:
- Notify your TA! As long as it is not contaminated with hazardous liquids, sweep it up. **Do not** pick it up with fingers! Dispose it in the broken glass container. Tongs should be used if hazardous liquids are present.
- Cuts:
- Notify your TA!
  - Minor cut: Wash the area with cold running water, dispense ointment onto a gauze pad, and then apply to the injury. Cover with a bandage. The TA will hand out the necessary first aid items.
  - Large cut: Apply pressure on the cut area and **call 911 immediately**!
- Burns:
- Notify your TA!
  - Minor: Hold the affected area under cold running water for 15 minutes.
  - Large: Same as for the minor burn and **call 911 immediately**!
- Fires:
- Notify your TA!
  - Small, on the bench or on the floor: Smother it and/or use water if it is NOT an electric fire.
  - Large, on the bench or on the floor: Use fire extinguisher. Aim it to the side of the fire and use sweeping motions across the base of the fire. DO NOT aim the extinguisher directly at the flames.
  - Fire extinguishers are located next to the doors in every General Chemistry Laboratory. If you are not familiar with the use of a fire extinguisher, take a moment to read the instructions.

  - Electric: Unplug appliances! Hit the red power button on the wall next to the prep room! **Do not** use water!
  - Person on fire: **Stop, drop, roll, and yell! Do not run! Do not** use a fire extinguisher on a person! Remove affected clothing, and get under the shower.
- Spills:
- Notify your TA!
  - Small, on bench or on the floor: Wipe it up with paper towel, and soak it with water. Concentrated acid or base spills first have to be neutralized with proper reagents.
  - Large spills on bench or on the floor: Call the TA and let the TA clean it up.
  - Small spills on self: Remove contaminated clothing and wash affected area for 15 minutes at least with running cold water.
  - Large spills on body: Call for help, and move away from the spill area. Remove contaminated clothing and rinse body in safety shower for at least 15 minutes with cold running water.

Emergency shower is located in the SW (210 and 204), SE (206), or NW (209) corner of the General Chemistry Laboratories.

- Splash on face: Call for help, and move away from spill area. Keep the goggles on and go to the eyewash station. Wash your face with the goggles on for 5 minutes. Then, take the goggles off and rinse your face with **eyes open** again for another 10 minutes with cold running water. Remove the contaminated clothing.

Eyewash stations are located next to the emergency showers in every lab.

- Fire alarm sounds:
  - Turn off appliances, gather your belongings, and leave the building through the stairs. Go to the grassy area and stay with your TA.

Always read the entire experiment write-up before coming to the laboratory and ALWAYS FOLLOW THE INSTRUCTIONS IN THE LAB MANUAL AND those GIVEN BY YOUR TA.

Laboratory Rules and Regulations

The three main rules for all undergraduate laboratories are as follows:

1. No student may work in a lab unless an instructor is on duty.
   (A TA and the lab director constitute the instructor for this course.)
2. Safety glasses must be worn at all times in the laboratory.
3. No eating, drinking, or smoking is permitted in the laboratory.
   (This also means that no food or drinks are to be brought in or stored in the laboratory.)

Attendance

During the academic term, a student is **required to complete all of the regularly scheduled labs**. You must attend the section that you are registered for or you will receive a grade of F for the lab course. Deductions will be made for arriving late to lab. If you miss the introductory lecture given by your TA, you will not be allowed to do the lab and will receive a zero for that lab. Provided you have completed all lab experiments, the lowest **nonzero** grade of an **entire experiment** will be eliminated at the end of the semester, and the remaining labs will count toward your final grade.

If You Miss a Lab

Missing a lab can happen; the problem is not unique. If you know you will be missing a lab prior to the event, such as representing the university at athletic contests, please contact your TA and the laboratory director as soon as possible, before the event, and arrangements will be made. In the case of an event happening that you are not in control of, such as car trouble, family emergencies, etc., it is imperative that you contact your TA and lab director during the timeframe of *that* lab, as all lab materials due that lab class are expected to be turned that day. There is no practical way we can repeat the various actual missed experiments once the apparatus and solutions have been put away. If a student misses a lab and does not follow the procedure stated above, the student will receive a zero for that lab. **Students who complete less than eight experiments will automatically receive a grade of "F" for this course.**

Requirements for Recording Data and Observations

There are five requirements for recording data and observations. All of these must be followed or the lab will not be accepted for grading.

1. Labs must be written in pen or will not be accepted for grading.
2. Data collected during lab time must be written in your lab report and must be signed by the TA **before** the student leaves the classroom. No lab will be graded without the TA's signature, and no data will be accepted that was not obtained during lab class.
3. Mistakes on raw data recordings should be crossed out and initialed by the TA.
4. No whiteout is allowed on **any** lab reports.
5. A large "X" must be placed in the unused space on any signed page by the TA.

Cheating and Honor Code

The honor code is strictly enforced. See FAC 6C5-4.001 **Honor Code—Academic Irregularities** (Florida laws and Florida Atlantic University administrative rules are subject to change at any time). See the current Undergraduate Catalogue for more complete details. Any kind of copying (or if you allow others to copy your work), plagiarism, falsifying data or using someone else's work is considered cheating, and will be handled as pursuant to the honor code. If students are found cheating on any portion of the lab report at any time, all involved students will receive, at minimum, a "0" as their grade for that lab assignment and, depending on the severity, can even receive an F in the course. Students will be notified of their action according to the Honor Code.

Grading

The course grade is determined by how well you performed the experiments in the laboratory. The course grade is not intended to reflect how good your intentions were, or how hard you worked. Lab reports are graded by the students' TA who follows a standardized grading rubric. The TA may lower the total score of each lab report with up to 5 points for major mishaps (classroom disruptions, not following procedure, attending class unprepared, etc.) during the lab. The lab director will assign final course grades based on grades and recommendations provided by the TA. Each lab grade will be based on, but is not limited to, the following criteria:

- ✓ Calculations—You **must** show calculation to receive **any** credit
- ✓ Proper use of significant figures and units is part of the grading
- ✓ Good percentage yields
- ✓ Quality of observation
- ✓ Quality of data
- ✓ Overall TA evaluation of your lab work
- ✓ Diligence and attention to detail
- ✓ Adherence to safety regulations
- ✓ Arrives to lab on time
- ✓ Coming to lab well prepared
- ✓ Timeliness of handing lab assignments—No late labs will be accepted
- ✓ Working independently on individual labs
- ✓ Working well in groups for the group discussion
- ✓ Completing experimental work and leaves the lab on time
- ✓ Handling equipment with knowledge and care

Change of Grade

If you disagree with any of the grades awarded to you, **an appeal must be made within one week of receiving the grade during lab**. The appeal should be made to your laboratory TA via e-mail first. If you are requesting a regrade, in that e-mail you must provide specifics as to why you would like the lab regraded. Once your TA responds, if you are still not satisfied, you may send an e-mail to make an appointment with the lab director to discuss your grade, within one week after your TA responds. **In that e-mail please make clear you have already e-mailed and spoken to your TA regarding that particular lab.** The lab director and your TAs will **not** entertain requests to regrade multiple labs. After two weeks have elapsed no change in grade may be requested by the student.

Grades will neither be curved nor rounded for this course. Grades will be available from the Registrar's office or on OASIS at the end of the semester.

LABORATORY EQUIPMENT

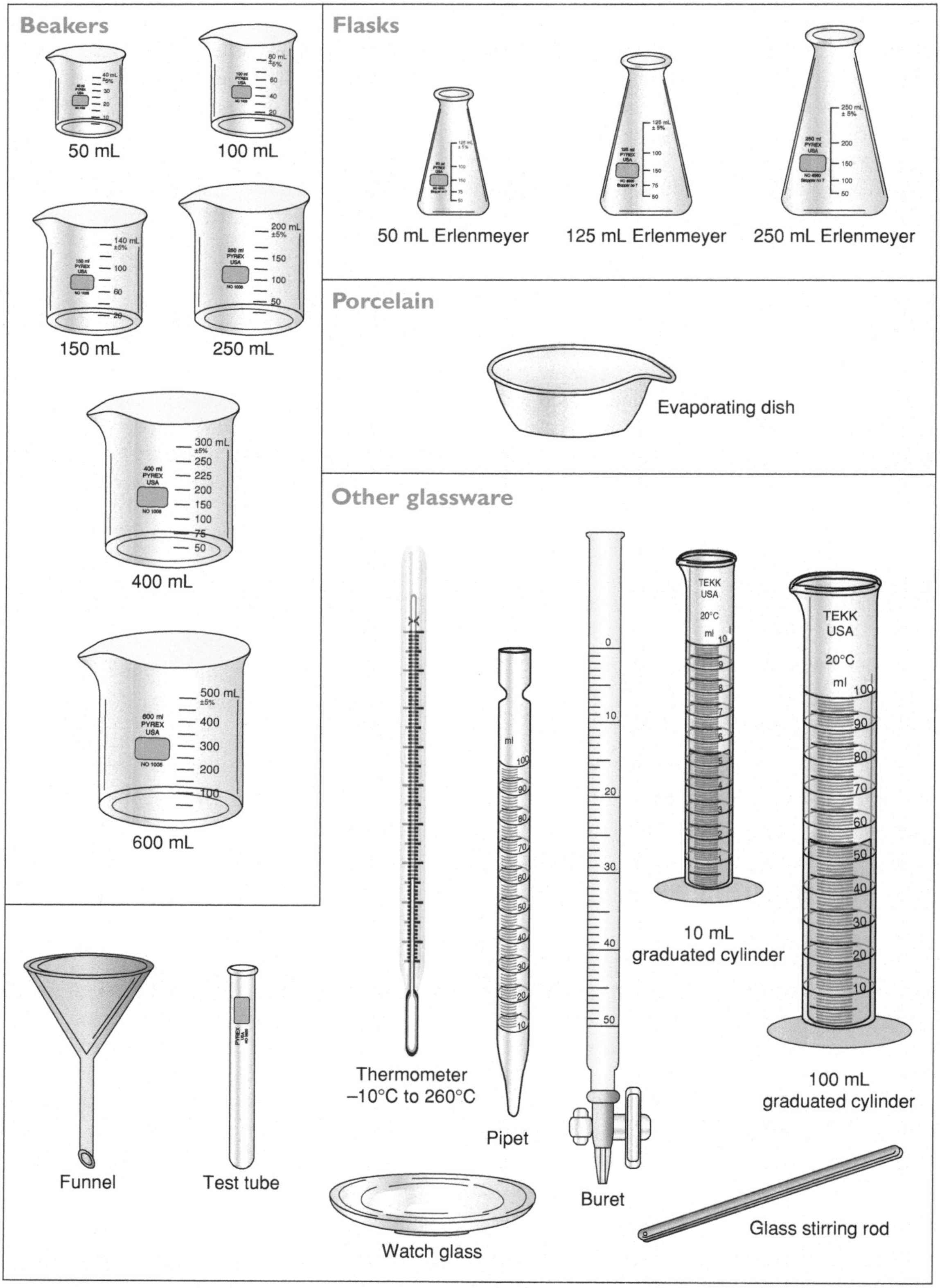

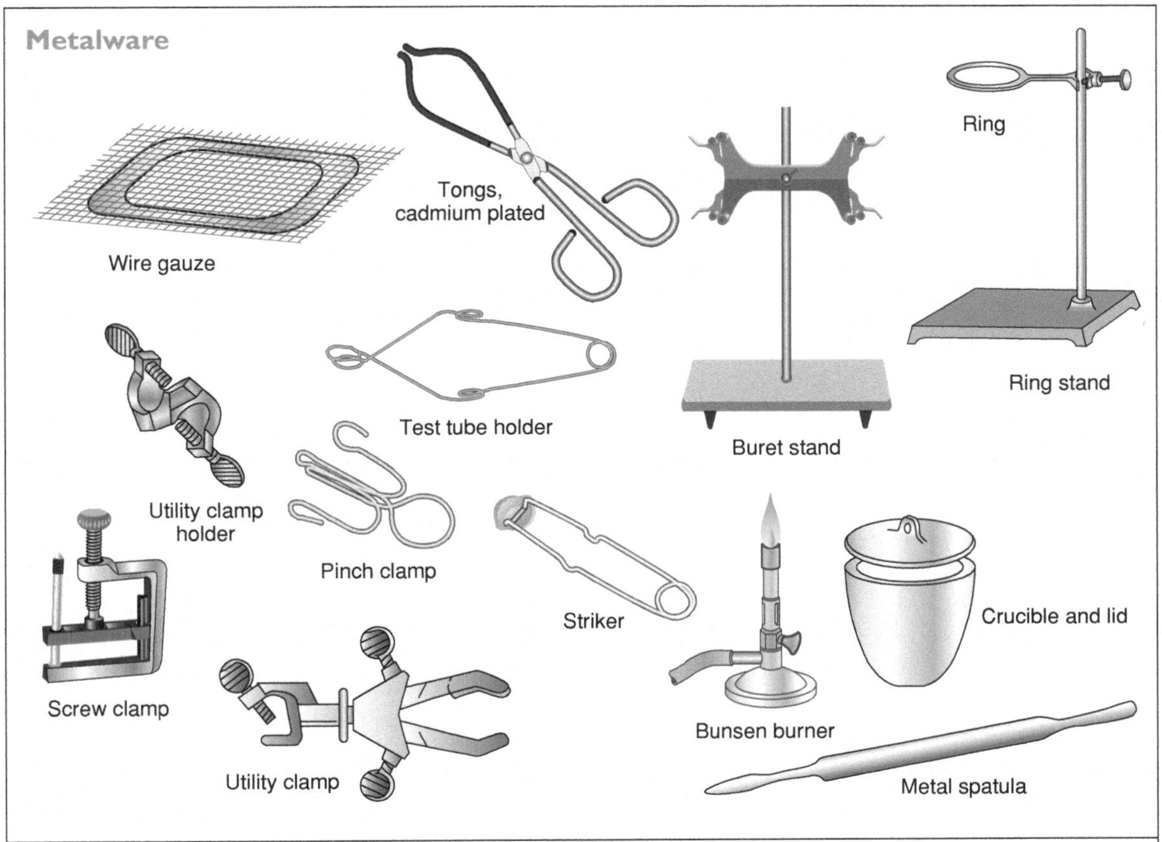
Metalware
Wire gauze
Tongs, cadmium plated
Ring
Ring stand
Buret stand
Test tube holder
Utility clamp holder
Pinch clamp
Screw clamp
Striker
Crucible and lid
Bunsen burner
Utility clamp
Metal spatula

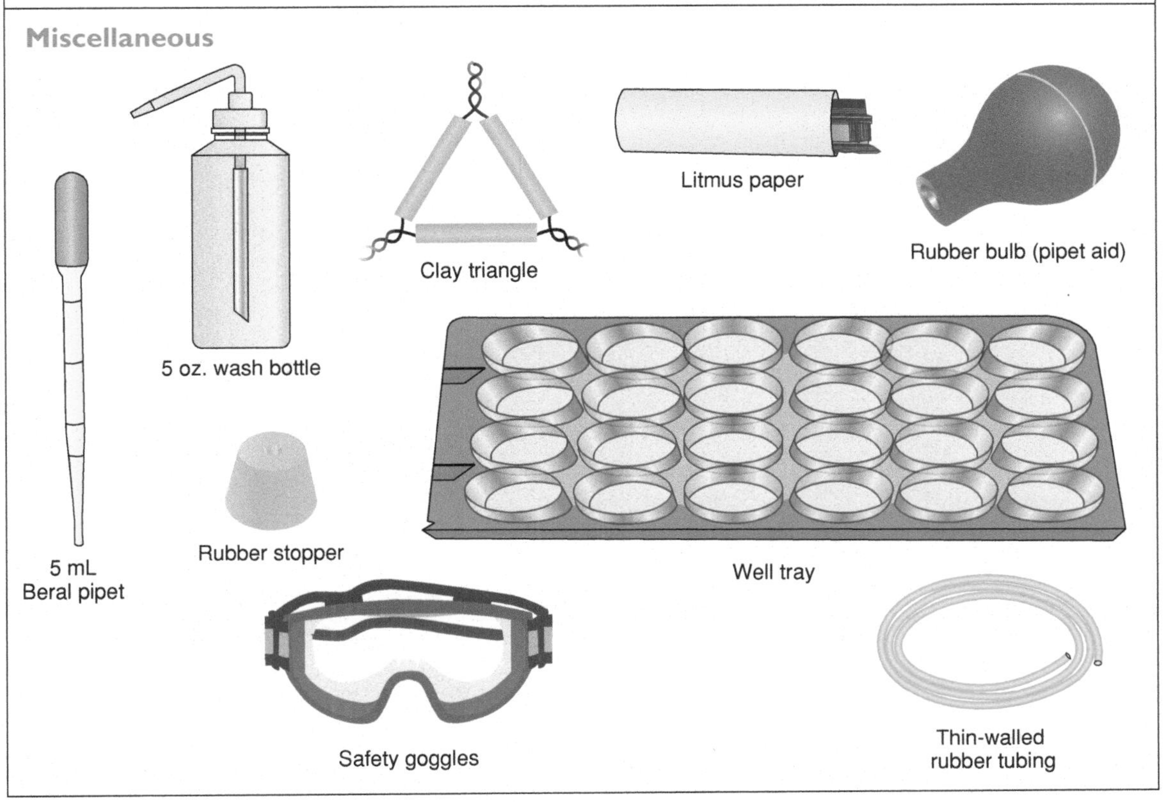
Miscellaneous
5 oz. wash bottle
Clay triangle
Litmus paper
Rubber bulb (pipet aid)
5 mL Beral pipet
Rubber stopper
Well tray
Safety goggles
Thin-walled rubber tubing

Drawer Inventory Sheet

Check your drawer to make sure that you have all of these items. If any of these items are missing from your drawer, notify your TA and obtain it to restock your drawer.

You should have one of each item, unless otherwise specified.

Complete your drawer inventory by putting a check mark for each item you have identified.

Glassware

____ 6 beakers (one of each: 600 mL, 400 mL, 250 mL, 150 mL, 100 mL, 50 mL)
____ 6 Erlenmeyer flasks (two 250 mL, three 125 mL, one 50 mL)
____ 2 evaporating dishes (porcelain)
____ long-stem glass funnel
____ powder funnel
____ 2 graduated cylinders (one 100 mL, one 10 mL)
____ 2 glass stirring rods
____ 10 test tubes (10 mL) in test tube rack
____ 2 watch glasses

Metal and Plastic Pieces

____ buret clamp
____ utility clamp with clamp holder
____ iron ring with clamp holder
____ Bunsen burner
____ striker
____ wire gauze
____ spatula
____ tongs
____ test tube clamp
____ 8 beral pipets
____ multi-well plate
____ clay triangle
____ pipet bulb
____ 2 rubber stoppers
____ wash bottle

The following items can be found on your laboratory bench:

____ ring stand
____ hot plate (shared with another student)

SAFETY QUIZ

1. According to the dress code, what should you wear in the laboratory?
   You must wear safety goggles/glasses, long pants, + close toed shoes. Long hair must be tied back. Lab coat, gloves, + aprons are used when needed.

2. What type of goggles should you wear and when should you wear them?
   Safety goggles/glasses that are department approved must be worn at all times. when in the lab.

3. In what cases might you be asked to leave the laboratory?
   If you are not dressed according to the dress code you may not stay in the lab.

4. Can you eat in the laboratory? Explain your answer.
   No food or drink is allowed in the lab, this is so that there is no contaminations.

5. What should you do if:

A. you broke a beaker and cut your finger?

- notify TA
- run cut under cold water + apply ointment / gauze (if minor)
- large cuts: apply pressure + call 911
- sweep up glass + put into glass disposal bin

B. chemicals have splashed on your face?

- notify TA
- call for help
- remove contaminated clothes
- rinse in shower (15 mins)

C. the fire alarm sounds?

- turn off appliances
- gather belongings
- leave building through stairs
- go to grassy area with TA

D. your lab manual has caught fire?

- notify TA
- smother + use water
- if larger use fire extinguisher

E. your shirt has caught fire?

- notify TA
- stop, drop, + roll
- do not use fire extinguisher
- remove clothing + then shower body

F. chemicals have spilled on your pants?

- notify TA
- remove contaminated pants
- shower with cold water (15 mins)

Safety Agreement

TA's copy

By signing this agreement, I hereby declare that **I understand and agree to follow all laboratory safety rules** established by the Department of Chemistry and Biochemistry at Florida Atlantic University.
I have watched the department's safety CD-ROM and have attended the safety lecture.

I understand that all safety rules are included in my lab manual for my information and they are accessible on the Internet at: http://www.science.fau.edu/chemistry/chemlab/General/safety.html

I understand that is my responsibility to make myself familiar with all safety regulations and to follow them at any given time when I am in the General Chemistry Laboratory.

I understand that I will have to successfully pass the Safety Quiz in order to be allowed to work in the General Chemistry Laboratories.

I UNDERSTAND THAT IN ANY ACCIDENT THAT IS CAUSED BY MY DISOBEDIENCE OF RULES, I AM SOLELY RESPONSIBLE FOR ALL CONSEQUENCES.

Cristina de Laski
Signature of student

8/24/22
Date

Cristina de Laski
Printed name of student

cdelaski2021@fau.edu
FAU e-mail address

Student Agreement

TA's copy

By signing this agreement, I hereby declare that **I understand it is my full responsibility** to do the following:

- Attend the section I am registered for, or I will receive an F grade for the lab course. Transfers must be made through the Registrar's Office.
- Read, understand, and adhere to the **syllabus** distributed at the first class session posted on Blackboard. fau.edu.
- Read, understand, and adhere to the **laboratory rules and regulations** printed in my laboratory manual.
- Keep track of the lab assignments and complete them in a timely fashion. Deadlines are to be met or I will accept the consequences of my actions (no late papers will be accepted).
- Make myself familiar with the experiments, and come fully prepared to all lab sessions.
- Make myself familiar with the use of Canvas, since the course is a Web-assisted course.
- View the appropriate technique from the video clips before each lab meeting.

I also understand it is my right to:

- visit **any** General Chemistry II TA's office hour to seek help.
- discuss my performance with my TA and lab director at any time throughout the term.
- contest any grade to the TA and lab director within the two-week period as described in the Grading policy, on page xi.

Cristina de Laski (signature)
Signature of student

8/24/22
Date

Cristina de Laski
Printed name of student

cdelaski2021@fau.edu
FAU e-mail address

Safety Agreement

Students's copy

By signing this agreement I hereby declare that **I understand and agree to follow all laboratory safety rules** established by the Department of Chemistry and Biochemistry at Florida Atlantic University.

I have watched the department's safety CD-ROM and have attended the safety lecture.

I understand that all safety rules are included in my lab manual for my information and they are accessible on the Internet at: http://www.science.fau.edu/chemistry/chemlab/General/safety.html

I understand that is my responsibility to make myself familiar with all safety regulations and to follow them at any given time when I am in the General Chemistry Laboratory.

I understand that I will have to successfully pass the Safety Quiz in order to be allowed to work in the General Chemistry Laboratories.

I understand that in any accident that is caused by my disobedience of rules, I am solely responsible for all consequences.

Cristina dL

Signature of student

8/24/22

Date

Cristina de Laski

Printed name of student

cdelaski2021@fau.edu

FAU e-mail address

STUDENT AGREEMENT

Students's copy

By signing this agreement, I hereby declare that **I understand it is my full responsibility** to do the following:

- Attend the section I am registered for, or I will receive an F grade for the lab course. Transfers must be made through the Registrar's Office.
- Read, understand, and adhere to the **syllabus** posted on Canvas.fau.edu.
- Read, understand, and adhere to the **laboratory rules and regulations** printed in my laboratory manual.
- Keep track of the lab assignments and complete them in a timely fashion. Deadlines are to be met or I will accept the consequences of my actions (no late papers will be accepted).
- Make myself familiar with the experiments, and come fully prepared to all lab sessions.
- Make myself familiar with the use of Canvas, since the course is a Web-assisted course.
- View the appropriate technique from the video clips before each lab meeting.

I also understand it is my right to:

- visit **any** General Chemistry II TA's office hour to seek help.
- discuss my performance with my TA and lab director at any time throughout the term.
- contest any grade to the TA and lab director within the two-week period as described in the Grading policy, on page xi.

Cristiani

Signature of student

8/24/22

Date

Cristina de Laski

Printed name of student

cdelaski2021@fau.edu

FAU e-mail address

G.C.2-1

Hess's Law

Objectives

By the end of this experiment you should be able to

- understand the process of heat transfer, with respect to enthalpy
- utilize and understand the process of calorimetry to determine heat transfer
- apply Hess's law to calculate enthalpy through direct and indirect processes
- determine the enthalpy of formation and enthalpy of reaction
- draw a graph using collected data

References

✓ Read Chapter 5 of your textbook

Pre-Lab Instruction

- Review the technique regarding the use of a balance

Background Information

Energy changes for chemical reactions can be determined by calorimetry, either directly or indirectly. Calorimetry is a procedure that allows us to measure the amount of energy (as heat given off or absorbed) under carefully controlled conditions. For today's experiment, we will be using a coffee-cup calorimeter to measure the energy change for certain reactions, by monitoring temperature changes of the system. To understand the process of energy transfer, we must explain the system and surroundings of our calorimeter. For this experiment, the system constitutes sample/substance only, no water, and water is part of the surroundings which include the cups and everything else. Since this is an insulated system, we make an assumption that no energy is transferred beyond the solution (system). When we perform a reaction in water, energy transfer can occur as heat change within the system. If exothermic, energy (heat) is transferred to the solution (system), and if endothermic, the opposite occurs. To measure the amount of energy (in the form of heat) absorbed or given off, we can use this equation:

$$\mathbf{q_{soln} = m_{soln} \times c_{solution} \times \Delta T_{soln}} \qquad \text{Eq. 1}$$

where q = heat, m = mass, c = specific heat capacity of water, and ΔT = change in temperature, in °C or K. We also make an assumption that since the solution is dilute, the specific heat capacity of water can be used (i.e., $c_{solution} = c_{water}$).

In an ideal system, and for this experiment, we will assume that none of the heat produced in the reaction is lost by the calorimeter to the rest of the universe (surroundings). With this assumption, the energy as heat lost by the reaction will equal the energy gained by the solution. We can represent this in the following equation:

$$\mathbf{q_{reaction} = -\,q_{solution}} \qquad \text{Eq. 2}$$

Combining these two equations, we can see that for the calorimeter

$$q_{reaction} = -m_{solution} \times c_{solution} \times \Delta T_{solution} \quad \text{Eq. 3}$$

This way we can calculate the heat change for the reaction (unknown) by determining the heat change for the solution (known). Since our reaction is being conducted at constant pressure, the heat change is given a symbol ΔH, called the **enthalpy change**. Different enthalpy changes for different types of reactions (such as formation and reaction) can be determined. Please refer to your text book for these.

For this experiment, using Vernier LabQuest 2 data collection interface and computer simulated software, you will monitor the temperature of the surroundings (water), and generate graphs from the collected " Temperature–Time" data. This will allow you to determine the initial temperature, $T_{initial}$, and the maximum (or minimum) temperature, T_{final}, that the Vernier device records. From this graph, $T_{initial}$ and T_{final} can be utilized to determine $\Delta T_{solution}$. If ΔT is positive, this means that heat was released to the solution and the reaction is exothermic. When q is calculated, then the value will be negative. The negative sign simply indicates the direction of heat flow, not the quantity. If ΔT is negative, heat was absorbed from the solution, the reaction is endothermic and q will be positive.

Once the heat quantity (q_{rxn}) is calculated, at constant pressure, $q_{rxn} = \Delta_{rxn}H$. We can then calculate $\Delta_{rxn}H$ per mole of substance, which can be demonstrated by the following equation:

$$\Delta_{rxn} H = q_{rxn} / mol_{substance} \quad \text{Eq. 4}$$

To measure the enthalpy change (ΔH) of a reaction directly is often difficult and sometimes impossible, depending on the reaction time, the pathway, or the lack of product stability. The enthalpy change of that reaction can however be determined indirectly from enthalpy changes for other reactions that can be measured. We can calculate the enthalpy change by using Hess' law, which states that if a reaction is the sum of two or more reactions, the enthalpy for the overall process is the sum of the enthalpy values of those reactions (Kotz, refer to 5.7 in your textbook).

According to Hess's law, you can measure the enthalpy of two reactions and add them to get the enthalpy of the total reaction, as in the following example.

Note: compounds found on both sides of the reaction cancel each other out.

$$C(s) + \tfrac{1}{2}O_2(g) \rightarrow CO(g) \qquad \Delta H_1 = -110.5 \text{ kJ/mol}$$

$$CO_{(g)} + \tfrac{1}{2}O_{2(g)} \rightarrow CO_{2(g)} \qquad \Delta H_2 = -283.0 \text{ kJ/mol}$$

$$C_{(s)} + O_{2(g)} \rightarrow CO_{2(g)} \qquad \Delta H_T = -393.5 \text{ kJ/mol}$$

In today's experiment, you will be determining the enthalpies of two separate reactions. In Part I of this experiment you will be determining the $\Delta H_{formation}$ of MgO and in Part II of this experiment you will be determining the $\Delta H_{reaction}$ of NH_4Cl.

Part I

For Part I, you will work in groups of two or three to determine the enthalpy change for the reactions of Mg and MgO with HCl to solve for $\Delta H_{formation}$ of MgO. You will start with Mg metal. One partner will measure the enthalpy change when Mg is added to 1.0 M HCl and measure the heat evolved or absorbed with a coffee-cup calorimeter (refer to 5.6 in your text to review calorimetry). The other partner will measure the enthalpy change that occurs when a measured amount of the MgO is added to 1.0 M HCl.

Once you determine the enthalpies of the two reactions, you can set up and solve the following reactions and apply Hess's law: (Note: equations are already balanced.)

$$Mg_{(s)} + 2HCl_{(aq)} \rightarrow MgCl_{2(aq)} + H_{2(g)} \qquad \Delta H_1 = ? \text{ kJ/mol}$$

$$MgO_{(s)} + 2HCl_{(aq)} \rightarrow MgCl_{2\,(aq)} + H_2O_{(l)} \qquad \Delta H_2 = ? \text{ kJ/mol}$$

To make these equations add up to the formation reaction of MgO, you will also need to include a basic equation with a published enthalpy. That equation is:

$$H_{2\,(g)} + \tfrac{1}{2}\,O_2\,(g) \rightarrow H_2O_{(l)} \qquad \Delta H_3 = -\,241.8\ \text{kJ/mol}$$

Utilizing Hess's law and manipulating these three equations and their respective enthalpy values, you can solve for $\Delta H_{formation}$ of MgO.

Part II

In Part II of today's experiment, you will also be working in groups of two or three to perform two tasks. For Task 1, you will measure the temperature change of two reactions:

$$HCl_{(aq)} + NaOH_{(aq)} \rightarrow NaCl + H_2O_{(l)} \qquad \Delta H_1 = ?\ \text{kJ/mol}$$

$$NaOH_{(aq)} + NH_4Cl_{(aq)} \rightarrow NH_{3(aq)} + H_2O_{(l)} + NaCl_{(aq)} \qquad \Delta H_2 = ?\ \text{kJ/mol}$$

One partner will measure the enthalpy change for the first reaction of 2.0 M NaOH and 2.0 M HCl, while the other partner measures the enthalpy change for the second reaction between 2.0 M NaOH and 2.0 M NH_4Cl. Using Hess's law, you can manipulate these equations to determine the $\Delta H_{reaction}$ of NH_4Cl from $NH_{3\,(aq)}$ and $HCl_{(aq)}$.

For Task 2, both students will work together to directly measure the enthalpy change for the reaction between $NH_{3(aq)}$ and HCl.

$$NH_{3(aq)} + HCl_{(aq)} \rightarrow NH_4Cl_{(aq)} \qquad \Delta H_3 = ?\ \text{kJ/mol}$$

This reaction, Reaction III, allows a direct, one-step reaction to determine the $\Delta H_{reaction}$ of NH_4Cl. This third reaction is performed to compare the results from determining the $\Delta H_{reaction}$ of NH_4Cl from both direct and indirect means.

Chemicals and Special Equipment

Vernier LabQuest 2 interface, computer, temperature probe, Styrofoam cups, plastic cover for the cups, 250 mL beaker, 50 mL or 100 mL graduated cylinders, Mg metal, MgO, 1.0 M HCl, 2.0 M HCl, 2.0 M NaOH, 2.0 M NH_3 (aq), and NH_4Cl

Hazard

H_2 gas, which is flammable, will be produced in the reactions, so use fume hood and avoid sparks. Metals and metal oxides are irritants; so wash hands before touching face. HCl is corrosive, highly toxic by ingestion or inhalation. NaOH is corrosive; skin burns are possible; dangerous to eyes. Ammonia liquid and vapor are extremely irritating to the eyes. Use the fume hood.

PROBE TOLERANCE: The stainless steel temperature probe should not be left in 2.0 M HCl solutions for more than ten minutes. This may ruin the probe. Therefore use of the temperature probe in this experiment must be strictly observed.

Procedure

One group of two students will be working on Part I of this experiment and a second group of two students will be working on Part II of this experiment. Depending on your assignment, read and follow the instructions below. At the end of the experiment, you must share the data you gathered with the other students on your lab bench (four students) to be able to perform all calculations.

Part I

If you were assigned Part I, you will be completing two reactions. Each group of two or three students will begin working with instructions for Reaction I with Mg metal and repeat the procedure for Reaction II with MgO. Make sure to record the system you worked on (Mg or MgO) on your data sheet. Obtain a LabQuest 2 interface device, two Styrofoam cups, a cover, and a sample of the Mg metal and MgO. The Styrofoam cups will serve as a calorimeter. You will use the two cups together to achieve greater insulation. Clean and dry your Styrofoam cups, place one into the other, and **weigh them together on the balance**. Record your data.

Magnesium Metal	Magnesium Oxide
0.150 g Mg	0.250 g MgO

Weigh your metal sample and record the exact weight of the metal on your data sheet. (Note: mass by difference is required). Cover the dish containing your metal sample with parafilm to avoid the sample escaping. **You will need to cut your Mg metal sample into small pieces to perform the experiment.** Fill a 400 mL beaker with 200 mL of deionized water for cleaning your probe and set aside for further use. Prepare a setup for calorimetric measurements as shown in Figure 1 :

(However your setup should include the cover).

Part I: Data Collection

Prepare your Vernier LabQuest 2 interface for data collection. (Note: there is an On/Off button.)

1. Press and release the power button located on the top left hand edge of the LabQuest 2 interface device. You should see the screen initializing, please wait. This may take a few minutes. When the file sensor screen appears, this indicates a successful power up.
2. Connect the stainless steel temperature probe to channel 1 of LabQuest 2 interface device. LabQuest App will auto-ID the connected sensor and automatically set the default collection rate for the sensor.
3. Ch1: Temperature will appear with the current temperature reading.
4. Click on "Rate" on the screen and change to 1 sample/second. Click "Done."
5. Click on the DURATION icon on the screen and change it to 180 seconds and click "Done."
6. You are now ready to collect data. Click on the File icon and go to "New" (discard old data if they appear on the screen).

Figure 1-1. Calorimeter

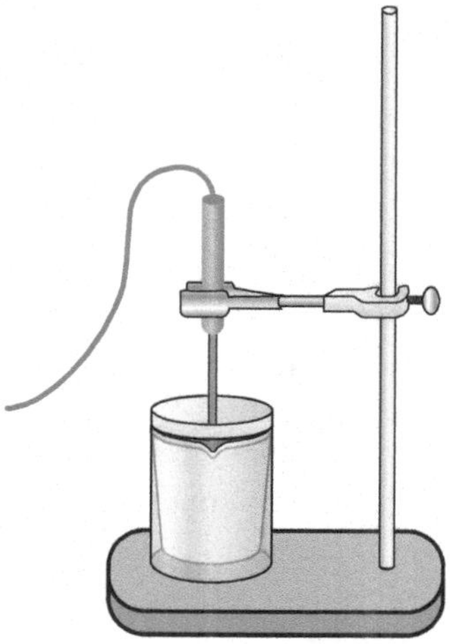

7. Click on the **green** triangle (Δ) on the left-hand bottom of the screen to begin data collection.
8. Data collection begins and LabQuest graphs the data in real time. It may take several seconds for the probe to equilibrate to the temperature of the solution.
9. After the designated time, press on the RED box in the lower left hand to stop the data collection.
10. To get the temperature min/max to display on the screen, tap analyze Statistics

Temperature

11. Rinse and dry the probe.
12. After completing your data collection, then connect the LabQuest 2 interface to the computer using the USB cable.
13. The TA will have the Logger Pro Program (software) on the computer open
14. Plug the LabQuest 2 USB into the computer; this will automatically transfer the graph into LoggerPro and the graph can be printed.
15. After the drivers have been installed, you will get a pop-up box that says something like "Remote data detected: retrieve data now?" Click "Yes."
16. Another pop-up box will appear asking if you want to open in the current file or an old one; click current OK.
17. Get the temperature max/min box by once again clicking Analyze then Statistics
18. Print
19. Once you have clicked on print or OK, a dialog box pops up which instructs you to disconnect Lab-Quest from the computer.
20. Once you have disconnected LabQuest from the computer, you are now ready for remote data collection. Return to the wet lab for data collection.
21. To turn off your Vernier Lab Quest 2, you must go to the home icon and push it to access the system screen. Once there you can push the shut down icon and turn off the lab quest.

Reaction I: Reaction between HCl with Mg(s)

Pour **exactly** 25.0 mL of 1.0 M HCl into your calorimeter and set up the apparatus as seen on page 4 of your lab manual. Place the temperature probe into the solution through the hole in your lid. (Note: Do not submerge the insulated portion of temperature probe below the solution you are measuring.) Press the Start/Stop button on the LabQuest interface to begin data collection. One student should be monitoring the time of the reaction. Measure and record the temperature of the acid at every second for 1.0 minute.

At 1.0 minute, add Mg solid into the cups quickly while constantly stirring it. Continue recording the temperature for another 2 minutes (120 seconds). Once data collection is complete hit the RED button to stop. Remove the probe from the solution and **immediately wash the stainless steel portion of the temperature probe in distilled water for approximately 2 minutes.** Dry the probe with a kim-wipe.

Measure the total mass of the cups plus the mixture and record your data. Dispose of your solution in the labeled waste container. Wash your cups out to prepare for the second part of this experiment.

Retrieve Data

To retrieve collected data, return to the computer room and connect the LabQuest 2 interface to a computer. Your interface MUST still have the temperature probe attached to it for proper retrieval. A "Remote Data Available" dialog box will pop indicating detection of remote data. Click on YES to retrieve the data and click on OK to retrieve the data into the current file. Logger Pro Software automatically generates a temperature–time graph from the data collected. Right click on the graph and select graph options. This allows you to put the appropriate title for your graph and other graphing options. Once you have titled your graph and appropriate axes, click "DONE." Then click on the statistics icon, , from the menu bar. This gives you the maximum and minimum temperatures from the graph, which need to be included on your graph. This information will allow you to calculate ΔT. The print options allow you to put your name, date, and page title on your graph. Using control P from your keyboard you may enter your name, date, and page title for your graph. Once your graph is printed, close the spreadsheet and do NOT save or make changes to the file. Unplug the USB from the computer and return to the lab.

Reaction II: Reaction between HCl with MgO

Repeat the entire procedure above for the MgO sample and record all data generated. Once the experiment is completed, remove the probe from the solution and **immediately wash the stainless steel portion of the temperature probe in distilled water for approximately 2 minutes.** Dry the probe with a kim-wipe.

Part II

If you were assigned Part II, you will also be working in groups of two or three for this experiment, performing three reactions. Reactions I and II utilizes Hess's law to determine the $\Delta H_{reaction}$ of NH_4Cl. Reaction III is a direct one-step reaction to determine the $\Delta H_{reaction}$ of NH_4Cl. You can compare $\Delta H_{reaction}$ of NH_4Cl solved from Reactions I and II to the value obtained from Reaction III.

Part II: Data Collection

Prepare your Vernier LabQuest 2 interface for data collection. (Note: there is an On/Off button.) Follow the steps in part 1 page 6.

You are now ready for data collection. Begin the wet lab for data collection.

Reaction I: Reaction between 2.0 M NaOH and 2.0 M HCl

Measure 50.0 mL of 2.0 M HCl in a graduated cylinder, transfer to a beaker and put aside for future use. Measure 50.0 mL of 2.0 M NaOH in the cleaned graduated cylinder and transfer into a second labeled beaker for future use. You are now ready to start your experiment.

Pour the 50.0 mL of 2.0 M HCl solution into the calorimeter. Lower the tip of the temperature probe into the HCl solution and click on the Start/Stop button. After 15 seconds, add the 50.0 mL of 2.0 M NaOH all at once and continue recording temperature values for 180 seconds. Click on the RED stop button. Once data collections are complete, rinse and dry the temperature probe in distilled water for 2 minutes and dispose of solutions down the drain for Reaction I (Reactions II and II will be collected in properly labeled waste containers).

Retrieve Data

You may retrieve your data by following the steps described in Part I. Record the initial and maximum temperatures in your data table.

Reaction II: Reaction between 2.0 M NaOH and 2.0 M NH_4Cl

Repeat the same procedure this time using 2.0 M NaOH and 2.0 M NH_4Cl. Please ensure you have the same parameters set for data collection. Pour exactly 50.0 mL of 2.0M NaOH solution into calorimeter. After 15 seconds, add 50.0 mL of 2.0M NH_4Cl all at once and continue recording the temperature.

Reaction III: Reaction between Solutions of 2.0 M HCl and 2.0 M NH_3 (aqueous)

Measure out exactly 50.0 mL of 2.0 M HCl and pour into calorimeter. The parameters for data collection remain the same. Lower the tip of temperature probe into solution and begin data collection. After 15 seconds, add 50.0 mL 2.0 M NH_3 solution (in fume hood) all at once and continue recording the temperature for 180 seconds. Repeat steps in Reaction I to retrieve your data. Record your initial and final temperatures in your data table. Dispose of solutions in the appropriate waste containers. **Once data collection is complete do not forget to rinse and dry the temperature probe. Make sure you turn off the LabQuest 2 by going to the home icon and pushing it to access the system screen. Then press the shut down icon to turn off the LabQuest 2. Return all components (LabQuest 2, remote, probe, and USB cable) to your teaching assistant for checkout.**

Students completing Part I of this experiment must share their data with students completing Part II of this experiment and vice versa. Both data tables must be filled out up to and including ΔT values prior to leaving the lab.

CALCULATIONS

Note: **The specific heat capacity of the water is 4.184 J/g · K.**

Part I

1. Calculate the mass of your metal by difference.
2. Calculate the molecular weight and then the number of moles of your metal and metal oxide.
3. Calculate ΔT: $\Delta T = T_{final} - T_{initial}$
4. Calculate the mass of the reaction mixture (Solid added to a solution) by difference.
5. Calculate q, the heat transferred with the metal and metal oxide.
6. Calculate $\Delta H_{reaction}$ per mole of Mg and MgO.
7. Utilizing the three equations provided and your knowledge of Hess's law, show how they can be manipulated to determine the equation and final value for $\Delta H_{formation}$ of MgO. Show balanced equations, cancelled out compounds, and enthalpies of all reactions as seen in the example in this lab.

Part II

1. Calculate ΔT: $\Delta T = T_{final} - T_{initial}$
2. Calculate the heat transferred q for each of the three reactions. Use 1.03 g/mL for the density of all solutions.
3. Calculate the number of moles of each reactant from the provided concentration (for Reaction 1, you are solving for moles of HCl; for Reaction II, you are solving for moles of NH_4Cl; and for Reaction III, you are solving for moles of NH_3).
4. Calculate $\Delta H_{reaction}$ per mole of HCl, NH_4 Cl, and NH_3, in kJ/mol.
5. Using the data collected for Reactions I and II, apply Hess's law to determine ΔH for NH_4Cl.
6. Using the equations provided from Reactions I and II, and your knowledge of Hess's law, show how they can be manipulated to determine the equation and final value for $\Delta H_{reaction}$ of NH_4Cl. Show balanced equations, cancelled-out compounds, and enthalpies of all reactions as seen in the example in this lab.

G. C. 2-1: LABORATORY REPORT

Considerations for Discussion and Conclusions Section

As you write your discussion and conclusions paragraphs in your lab notebook, here are some tips and questions you should be answering. (This is in addition to the "Guidelines for a good discussion and conclusion" handout provided to you at your first lab class.)

- What are major concepts utilized in this experiment?
- What special instrumentation was relevant to the completion of this experiment and why was it used as it pertains to the calculation of heat quantity?
- What did we assume about heat lost and the calorimeter that allowed us to determine ΔT?
- From the drawn graphs and our ΔT values, can we determine additional information about heat transfer (exothermic versus endothermic)?
- How were the equations used in relation to Hess's law for both Parts I and II to solve for ΔH?
- For Sources of Error:
  - Part I: The theoretical values for $\Delta H^0_{formation}$ of MgO is -601.8 kJ/mol.
  - Determine your percentage error for this experiment.
- For Part II: Compare the value for $\Delta H_{reaction}$ for Reaction III that you obtained in this experiment with the calculated value of Reaction III using Hess's law. How do these values compare?

G. C. 2-1: LABORATORY REPORT

Name: ______________________________

Lab partners' names: ______________________________

YOU MUST SHOW ALL CALCULATIONS TO RECEIVE CREDIT FOR THEM!

Proper use of significant figures and units is part of the grading.

Part I: Determination of $\Delta H_{formation}$ of MgO

Data	Magnesium	Magnesium Oxide
Mass of weigh dish		
Mass of dish + sample		
Mass of sample		
Molecular weight of Mg/MgO		
Number of moles of Mg/MgO		
Mass of Styrofoam cups		
Mass of cup + reaction mixture		
Mass of reaction mixture		
$T_{initial}$ (from graph)		
T_{final} (from graph)		
ΔT (from graph)		
q, heat transferred		
$\Delta H_{reaction}$		
$\Delta H_{formation}$ of MgO		

G. C. 2-1: LABORATORY REPORT

Name ____

Lab partners' names: ______________________________________

YOU MUST SHOW ALL YOUR CALCULATIONS TO RECEIVE CREDIT FOR THEM!

Proper use of significant figures and units is part of the grading.

Part II

Data	Reaction I	Reaction II	Reaction III
$T_{initial}$ (from graph)			
T_{final} (from graph)			
ΔT (from graph)			
q, heat transferred			
Number of moles of reactant			
$\Delta H_{reaction}$			
ΔH reaction NH_4Cl (calculated from Reactions I and II)			

Students must share data with entire group before they leave the lab. All students should have data for both Parts I and II. **You will be required to complete ALL calculations associated with both Parts I and II.**

G.C.2-2

Investigation of the Relationship between Volume and Temperature

Objectives

By the end of this experiment you should be able to

- identify the relationship between volume and temperature
- investigate some properties of a gas
- use a Bunsen burner
- convert pressure and temperature values of different units

References

✓ Read Chapters 10.2–10.4 in your textbook.

Pre-Lab Instruction

- Review the techniques regarding the use of a Bunsen burner and the leveling of the flask in the ice bath.

Background Information

In today's experiment, you will be investigating the relationship between volume and temperature of a gas held at constant pressure. In 1802, this relationship was experimentally proven and published by Joseph Gay-Lussac, fifteen years after the French scientist Jacques Charles made the discovery. This law is known today as Charles' or Gay-Lussac's law; details about this law can be seen in Chapter 10.2 of your textbook. We will attempt to investigate this relationship by finding the volume-to-temperature ratios of a gas under three different temperature conditions. The gas we will be investigating is air, and the volume that hot air, air at room temperature, and cold air occupies in an Erlenmeyer flask will be determined under hot temperature, room temperature, or cold temperature conditions, respectively.

To quantify the volume of gas under hot (boiling) temperature conditions, we will have to assume the maximum volume of the Erlenmeyer flask is equal to the volume of the hot dry air, i.e. all the space in the Erlenmeyer flask is occupied by the gas. To quantify the volume of a gas under room or cold (ice) temperature conditions, however, we have to submerge the flask containing the hot air into a water bath at each temperature and determine how much water enters the flask to displace the space created by the cooling air. Since the hot air will be in contact with water, you will have to account for any water vapor being formed. This is important when performing calculations to determine the volume of the cold air and the volume of air at room temperature.

You will be working in groups of two for this experiment; each student will determine volume of air at boiling temperatures and either room or cold temperatures. Two trials will be performed to show consistent data. Once you have completed both the trials, you will combine data with your partner and draw a graph of all the tested values to determine the relationship between volume and temperature.

Chemicals and Special Equipment

Ice, glass and rubber tubing with pinch clamp or screw clamp, acetone

Hazard

Inspect glassware for cracks and chips before use. Be careful when transferring hot glassware. Acetone is both volatile and flammable. Please handle with caution and use in well-ventilated area.

Procedure

You will be working with a lab partner for this experiment. Each of you will be assigned one system, either the ice system or the room temperature system along with the boiling water system. You will perform two trials, one after the other. Obtain a thermometer and an apparatus consisting of a piece of glass tubing, rubber tubing, a rubber stopper, and a pinch clamp or screw clamp. Your teaching assistant will set up plastic tubs for the ice water bath and for the room temperature water bath in the lab. Depending on the system assigned, follow the subsequent instructions and then exchange results with your partner.

Clean two 125 mL Erlenmeyer flasks and dry the inside of them completely with paper towels or use acetone to rinse them out. Make sure the acetone is disposed of properly in a waste container. Firmly place the stopper with the tubing into the Erlenmeyer flask and **with a marker, draw a line on the flask at the bottom of the stopper.** This will assist in determining the maximum space the air can occupy in your flask. Set up the apparatus as shown in Figure 2-1. Attach a utility clamp to your ring stand. Place the beaker with flask on your hot plate as shown.

Figure 2-1. Setup for investigation

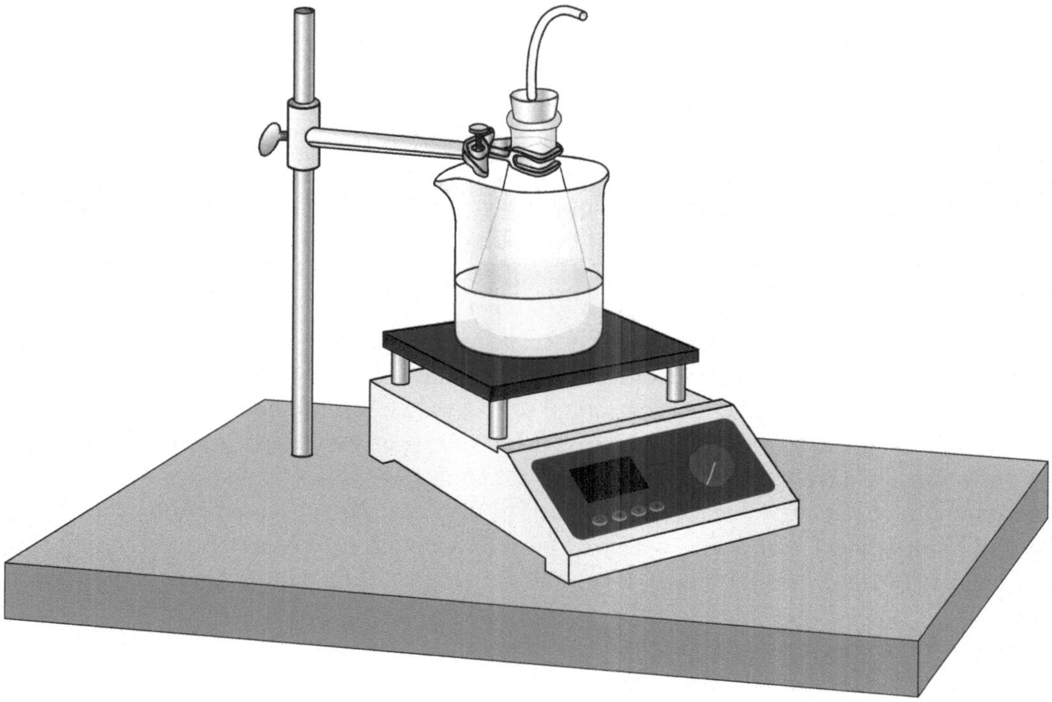

Fill up your 600 mL beaker halfway with water and put 2–3 boiling chips into it. Place this water bath on the hot plate. Attach the flask to the utility clamp and submerge it into the water as much as possible. Do not allow the flask to touch the sides or the bottom of the beaker. Remove the pinch clamp from the tubing to avoid explosive pressure building in the flask. **Caution: Make sure all utility clamps are tight, the beaker is balanced, and set up is perfect before you begin heating!** Boiling water can cause serious burns! Turn on the hot plate and start heating your water bath. Bring your water bath to a boil and heat the flask for 6–10 minutes. At this point you may have to add more water to the beaker to keep the water level of the bath constant.

While avoiding contact with the glass sides, place the thermometer into the water bath next to the Erlenmeyer flask for 3–4 minutes and record the temperature of the water on your data sheet. Carefully close the pinch clamp tightly as close to the glass tubing as possible. With the utility clamp, immediately remove the flask from the boiling water.

Place the flask **into the bath for your assigned system** (either ice bath or room temperature water bath). Holding the utility clamp to avoid frozen fingers, submerge the flask and tubing **completely** into your assigned water bath and open the pinch clamp. Keep the flask in that position for 5 minutes.

After 5 minutes, begin to raise the flask in the same position until the water level inside of the flask is the same as the water level in the bucket, **keeping the tubing still submerged**. This allows for the same air pressure to be inside of the flask as it is outside. Once the water is leveled, tightly close the pinch clamp in the same position as it was before. See Figure 2-2. Remove the flask from your assigned water bath making sure you can still see the marker line drawn on the flask at the bottom of the stopper. Remove the stopper from the flask. Measure and record the volume of the water in the flask by pouring it into a 100 mL graduated cylinder. Measure and record the temperature of your assigned water bath.

Determine the maximum volume of the Erlenmeyer flask, **which is equivalent to the volume of the hot dry air**. To do this, fill up the Erlenmeyer flask completely with water. Then insert the stopper again making sure it is aligned to the original mark, while keeping the pinch clamp open. Close the clamp and carefully turn the flask upside down, thus allowing the extra water to drain out of the tubing.

Remove the stopper and measure the volume of the water in the flask. Do this by emptying the flask into the **dry** 100 mL graduated cylinder. Keep in mind that the volume of the water will be more than 100 mL, thus you will have to empty and refill your graduated cylinder, then add up the partial volumes. You should not exceed the 100 mL mark on your graduated cylinder.

Obtain the barometric pressure data from your teaching assistant. Once you have obtained the data from your lab partner, draw a graph on the computer that will assist you in determining the relationship between volume and temperature. **(Hint: you will need three plotting points on the graph to correspond to the three temperatures and the three volumes of dry air determined.)**

Figure 2-2. Submerging the flask in the water bath

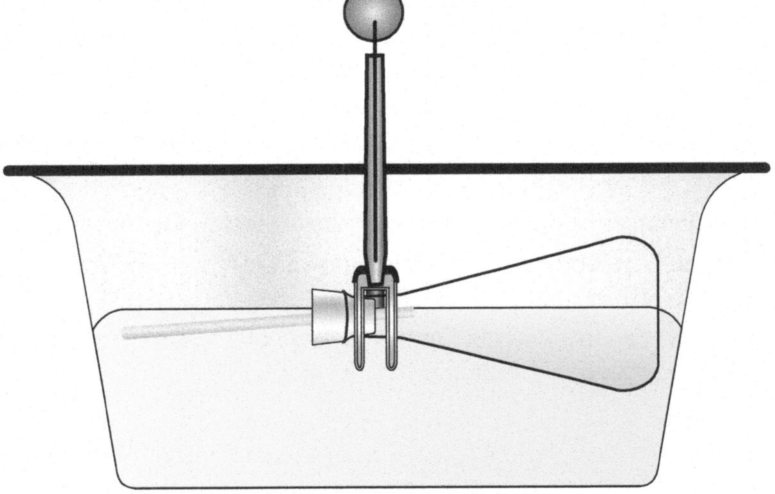

Calculations, Conversions, and Values

1. **Conversion to torr**: 1 mm of Hg = 1 torr
2. Vapor pressure of water at different temperature conditions: **Refer to Appendix G (p. A-22) in your text.**
3. Pressure of the "assigned system," dry air = barometric pressure (torr) − vapor pressure of water at the measured temperature of the system's water bath (torr).
4. Volume of the "assigned system," wet air = volume of hot dry air − volume of water in the flask.
5. Volume of the "assigned system," dry air (to adjust for the vapor pressure) = (volume of "system", wet air) × (pressure of "system," dry air / barometric pressure).
6. Graph data using correct calculated values and units.

Significant Figures

Significant figures are the correct number of digits that reflects the measurement performed. It is directly linked to the instrument or glassware used to make the measure. Significant figures rules are always followed when reporting a measure.

- **Tolerance or calibration of the instrument used**
  - 10.0 mL graduated cylinder: The smallest division of this graduated cylinder is 0.2 mL; therefore our reading error would be 1/100 of the smallest division. An appropriate reading of this volume would be 4.65 mL.
  - Electronic balance: The exact mass of the sample as seen on the screen of the electronic balance should be given. An appropriate reading of this mass would be 5.592 g.
- **Zeros in a number**
  - Leading zeros are never significant. Change your data to scientific notation to express the correct amount of significant figures when needed (i.e., there are four significant figures in the following numbers 03.205, .03205, and 0.003205).
  - Imbedded zeros are always significant (i.e., there are four significant figures in 3.025 and 3.205).
  - Zeros after a number are significant only after the decimal point is been specified (i.e., there are four significant figures in 0.003250).
- **Addition and Subtraction**
  - The amount of significant figures to be reported after a mathematical operation of this type is determined by adding or subtracting the number that has the **least amount of digits after the decimal point.**
- **Multiplication and Division**
  - The amount of significant figures to be reported after a mathematical operation of this type is determined by multiplying or dividing the number that has the **least amount of significant figures.**
- **Rounding rules**
  - Always round up to the nearest number reflecting the correct amount of significant figures.
  - If the number to round is an odd number and is followed by a number 5, then round up (i.e., 3.2055 is round up to 3.206 to reflect only four significant figures).
  - If the number to round is an even number and is followed by a number 5, then let it be (i.e., 3.2045 is round up to 3.204 to reflect only four significant figures).

Graphing Rules

- A graph has to have a title. This title need three basic elements: type of graph, experiment or concept tested in the laboratory, substance or material under study.
- A graph should be large. It should cover at least ⅔–¾ of the page.
- Data points should be clearly marked and spread over "most" of the graph. Do not cluster data points to a small part of the graph paper.
- A graph has to have clearly labeled and consistently divided axes.
- A "best fit curve" should be fitted to the data points. This means you should draw a single, **straight line** in between or touching most of the points. You might have to discard one or two data points if they are off. Make sure your graph goes through the (0,0) points.
- A slope of the best fit line should be determined by rise over run formula **($Y_2 - Y_1 / X_2 - X_1$). Do not take by default a pair of coordinates from your data set unless they are over the best fit line.**
- Always show, in a distinctive way, the pair of coordinates that you use to calculate your slope of the graph.

Graphing Using Excel 2007

To enter data

- In column A, row 1: enter the name of the independent variable.
- In column B, row 2: enter the name of the dependent variable.
- Enter data points (x, y) in the columns/rows below the titles you entered.

To graph the data points

- Highlight data points and titles. Select INSERT: SCATTER (top left icon).
- Click LAYOUT: AXIS TITLES, for x axis click TITLE BELOW AXIS, for y axis click ROTATE TITLE.
- To change graph title, double click on text or right click on text and select edit text.

To draw the linear best fit line (This Does Not Apply to pH Curves)

- Click LAYOUT: TRENDLINE: MORE TRENDLINE OPTIONS: LINEAR and DISPL:AY EQUATION ON CHART

G. C. 2-2: LABORATORY REPORT

CONSIDERATIONS FOR DISCUSSION AND CONCLUSIONS SECTION:

As you write your discussion and conclusions paragraphs in your lab notebook, here are some tips and questions you should be answering. (This is in addition to the "Guidelines for a good discussion and conclusion" handout provided to you at your first lab class.)

- What is your experimental relationship between temperature and volume? Was it direct or indirect? How/why did you determine this? (i.e., using your graph drawn).
- Does the relationship between volume and temperature you determined comply with discoveries made by Charles? Explain why or why not.
- Why did you have to calculate volume of dry, cold air instead of using the obtained volume of wet, cold air?
- Using your understanding of Charles' law and the experimental data obtained, explain what is happening to the gas at a molecular level when there is a change in temperature.

G. C. 2-2: LABORATORY REPORT

Name: ____

Lab partner: ______________________________ **Assigned system** ______________________

YOU MUST SHOW ALL YOUR CALCULATIONS TO RECEIVE CREDIT FOR THEM!

Proper use of significant figures and units is part of the grading.

Measurements	Trial No. 1	Trial No. 2
Temperature of ________ water bath, °C Room Temp/Cold*		
Temperature of ________ water bath, K		
Volume of water in the flask		
Temperature of hot water bath, °C		
Temperature of hot water bath, K		
Volume of hot dry air		
Average volume of hot dry air		
Barometric pressure, mm of Hg (From TA)		
Barometric pressure; torr		
Pressure of ________ dry air, torr		
Volume of __________, wet air		
Volume of ___________, dry air		
Average volume of _________, dry air		
Temperature of ________ water bath, **from lab partner**		
Average volume of _____ dry air, **from lab partner**		
Circle your system: Room Temp (RT) Cold		

G.C.2-3

Analysis of Cations

Objectives

By the end of this experiment you should be able to

- understand solubility and precipitation and how it applies to the equilibrium constant, K_{sp}
- use analysis to identify metal cations in an unknown sample based on experimentation
- perform a flame test

References

✓ Review Chapter 17.4–17.7 in your text book.

Pre-Lab Instruction

- Review the technique regarding the use of a Bunsen burner.

Background Information

Analysis is an investigation of the characteristics of a substance to discover its identity. Chemists use analysis to determine the components of a sample. During this experiment, you will be asked to design a procedure with which you can perform an analysis of a liquid sample that contains three different cations. Since it is a small-scale experiment, you will use small amounts of chemicals, such as drops, and specific equipment.

When we carry out reactions of various cations with a certain set of reactants, some will form solid precipitates and some will not. Reactions that involve precipitation occur when one of the products is insoluble in water. Even though the compound precipitates, a very small amount of the compound will dissolve in water. An equilibrium is established where the ions of the insoluble salt continuously go in and out of solution. The degree to which an insoluble compound can dissolve can be expressed in terms of K_{sp}, the solubility product constant. K_{sp} has already been introduced to you in the previous lab experiment. For this experiment, you may compare K_{sp} values of different compounds to predict whether they will precipitate or not. Note: K_{sp} values can only be compared for compounds that contain ions with the same stoichiometric ratio (see p. 667 for details). Larger solubility (K_{sp}) values indicate greater solubility and hence less precipitation. Table 3-1 provides a list of K_{sp} values that may be of importance.

Table 3-1: Some insoluble compounds and their K_{sp} values.

Formula	Name	K_{sp} (25°C)
$BaSO_4$	Barium sulfate	1.08×10^{-10}
$PbSO_4$	Lead sulfate	2.53×10^{-8}
$CaSO_4$	Calcium sulfate	4.93×10^{-5}
$Cu(OH)_2$	Copper hydroxide	4.80×10^{-20}
$Co(OH)_2$	Cobalt hydroxide	1.00×10^{-15}
$Ni(OH)_2$	Nickel hydroxide	6.00×10^{-16}
$Ca(OH)_2$	Calcium hydroxide	6.50×10^{-6}

The analytical scheme will be based on the insolubility of specific precipitates formed between the cations and reagent chemicals. These differences will help you identify the ions in your unknown sample. Reactions will be carried out in solution; thus it will be easy to recognize whether a solid precipitate has formed or not.

The list of the possible cations, the chemicals that you need to perform your analysis, and a table of the possible reactions that can take place are provided to you. Your unknown sample can contain only one cation from each group.

Cations can be categorized into groups based on their reaction with certain reagent chemicals. We differentiate between four groups. Ions within the same group will give reactions with the same chemical. For example, ions in Group 1 will react with HCl, producing a precipitate. This is not true for any ions in the other three groups.

The analytical scheme will be based on the reactions of the ions to the "group reagent" as well as to specific reagent chemicals. Table 2 shows which ions belong to the same group. As a first step, you will identify which group your unknown ions belong to and then you will identify the ions you have within each group.

Sometimes, reaction with the group reagent itself will enable you to identify your unknown ion. But most of the times you will have to perform further testing, carrying out characteristic tests that are specific for a particular ion. **For this lab, please perform all the characteristic tests indicated in Parts I and II of this experiment, as obtaining negative results to a test should also be used to correctly identify the cations present in your unknown sample**. For Part II of your procedure, the following techniques can be employed:

1. Reacting ions with a reagent should be carried out in a multi-well plate.
2. Flame test: This can help identify certain ions. Ions with electron configurations allowing for the outermost electron to reach an excited state easily can be identified by flame test. The energy transferred by the flame is sufficient to excite an electron and as the electron drops from the excited state, it changes the color of the flame. The color is characteristic for a particular ion.
3. In this experiment, Pb^{2+} will not be present in the same solution in which Ba^{2+} or Ca^{2+} might exist. Students could use this as evidence to help identify ions in their unknowns.

Performing a Flame Test—DONE BY TA ONLY!

- Please review the use of a Bunsen burner from the video clips. Light your Bunsen burner. Have your sample ready in a clean test tube.
- For the flame test, you will use a special wire called nichrome wire. The wire is inserted into a cork and has a loop at the end.
- Clean the loop of the wire by dipping the loop in 6.0 M HCl solution, then heat the wire for 15 seconds and rinse with deionized water. Repeat this procedure.

Figure 3-1. Performing a flame test

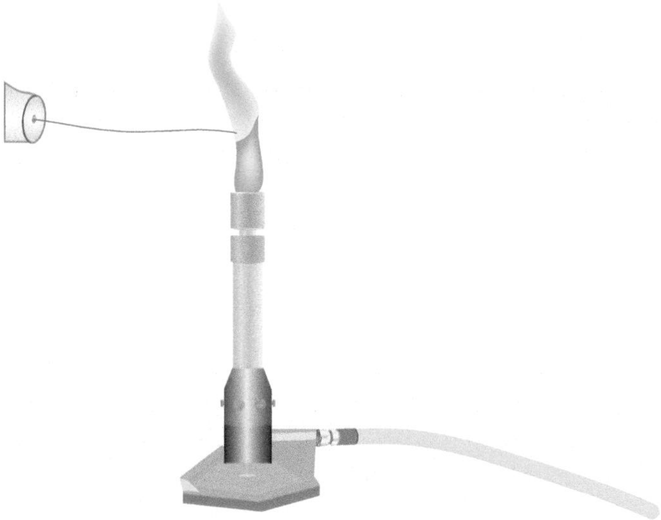

- Heat the loop of the wire until it glows red.
- Dip the wire into the solution you would like to test. The loop will collect and hold some of the solution. Hold the wire into the side of the flame and observe the color behind the cobalt glass provided.
- Repeat the flame test without the cobalt glass, and from the color of the flame, identify your ion.

CHEMICALS AND SPECIAL EQUIPMENT

Unknown sample contains three ions, at most one from each of the following sets. No unknown sample will contain two ions from the same set.

SET 1: Pb^{2+}, Ag^{+}
SET 2: Fe^{3+}, Cu^{2+}, Ni^{2+}, Co^{2+}
SET 3: Ba^{2+}, Ca^{2+}
SET 4: K^{+}, Li^{+}

Chemicals to identify sets: 6.0 M HCl, 0.10 M Na_2CO_3, and 1.0 M NaOH
Reagent chemicals to identify individual ions: 0.10 M HCl, 0.10 M K_2CrO_4, 0.10 M KSCN, 0.10 M KI, 0.10 M $K_4[Fe (CN)_6]$, and 1% dimethylglyoxime (DMG) solutions, 6.0 M HCl for flame test, and 0.1 M Na_2SO_4
Equipment: Nichrome wire with cobalt glass for flame test, multi-well plate, test tubes, Beral pipets, and toothpicks

Waste Collection

Waste collection in pre-labeled containers: Ions that have been mixed with a reagent chemical have to be collected separately, in individual containers.

HAZARD

Nitrate solutions are oxidizers and irritants. Avoid skin contact. HCl is a strong acid and NaOH is a strong base. Handle them carefully! Wash your skin immediately if they get on your skin, and notify your TA.

PROCEDURE

1. Carefully clean a test tube. Any kind of contamination can lead you to obtain false results.
2. Obtain your unknown sample. Your sample will be about 5.0 mL. You will have to work wisely so that the amount will be sufficient for you to complete your investigation.
3. Your task is to identify which sets and, within those sets, which ions are present in your unknown sample.
4. To complete your task, you will perform a series of precipitation reactions and carry out several flame tests.
5. Before you start, create a flow chart to plan your experimental procedure. A sample is shown below.

Figure 3-2. Flow chart for a procedure

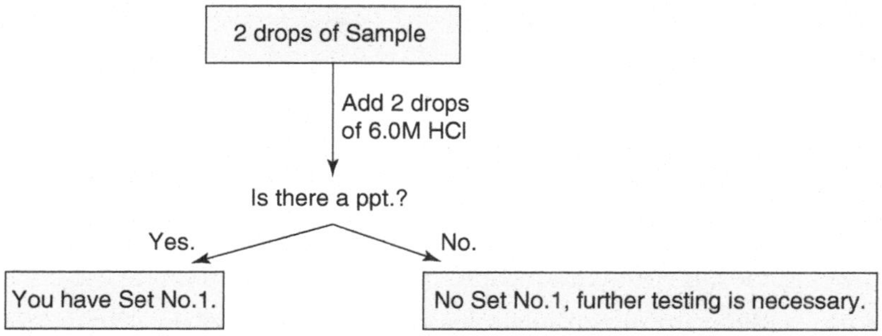

Part I. Identify the Sets in Your Unknown

1. The following reactions occur between ions of the same set and the provided chemicals:
   - Set 1. Ions (Pb^{2+}, Ag^{+}) will form precipitates with HCl (test with 6.0 M solution).
   - Set 2. Ions (Fe^{3+}, Cu^{2+}, Co^{2+}, Ni^{2+}) will form precipitates with NaOH.
   - Set 3. Ions (Ba^{2+}, Ca^{2+}) will form precipitates with Na_2CO_3.
   - Set 4. Ions (K^{+}, Li^{+}) will not form precipitates with any of the chemicals.

   The possible reactions are given in Table 3-2.
2. When performing a precipitation reaction, use drops only. Place 2–3 drops of your unknown sample into a well of the multi-well plate and then add 2–3 drops of the appropriate reagent; use a toothpick to mix the solutions.
3. Write down every reaction performed and your results. Identify the three sets present in your unknown.

Part II. Identify the Three Cations Present in Your Unknown Sample

4. Once you have identified the three sets, use the following information to distinguish between the ions within the sets.
5. Perform ALL reactions to test for the ions you suspect might be present in your unknown sample.
6. Follow your TA's instruction on proper waste disposal.
7. Write down every performed reaction and the results.

Table 3-2: Observed reactions to identify the sets

Set No.	Ions	6.0 M HCl	1.0 M NaOH	0.1 M Na_2CO_3
1	Pb^{2+}	White ppt.	—	White ppt.
	Ag^{+}	White ppt.	—	White ppt.
2	Fe^{3+}	—	Brown ppt.	—
	Cu^{2+}	—	Blue ppt.	—
	Co^{2+}	—	Green ppt. turns black	—
	Ni^{2+}	—	Green ppt.	—
3	Ba^{2+}	—	—	White ppt.
	Ca^{2+}	—	—	—
4	K^{+}	—	—	—
	Li^{+}	—	—	—

Note: For the reactions above, a "—" indicates that no precipitate can be seen.

Table 3-3: Observed reactions to identify the ions

Set No.	Ions	Specific Reactions	Flame Test
1	Pb^{2+}	+ K_2CrO_4 = yellow ppt. + KI = yellow ppt. + 0.10 M Na_2SO_4 = white ppt.	NA
	Ag^{+}	+ KSCN = white ppt. + K_2CrO_4 = red/brown ppt. + 0.10 M HCl = white ppt. + KI = light yellow ppt.	NA
2	Fe^{3+}	+ KSCN = deep red soln. eventually turns to white PPT + $K_4[Fe(CN)_6]$ = blue ppt.	NA
	Cu^{2+}	+ $K_4[Fe(CN)_6]$ = red ppt. + KSCN = no ppt.	Blue
	Co^{2+}	$K_4[Fe(CN)_6]$ = grass green ppt.	NA
	Ni^{2}	+ DMG = red/pink ppt. + $K_4[Fe(CN)_6]$ = green ppt.	NA
3	Ba^{2+}	+ K_2CrO_4 = yellow ppt. + 0.1 M Na_2SO_4 = white ppt.	NA
	Ca^{2+}	+ 0.10 M Na_2SO_4 = no ppt. + 1.00 M Na_2SO_4 = cloudy white ppt. with stirring (mix it thoroughly for 30 seconds or you might not be able to observe any reaction)	Orange-red
4	K^{+}	NA	Light purple
	Li^{+}	NA	Bright red

G. C. 2-3: LABORATORY REPORT

PART I. IDENTIFY THE SETS IN YOUR UNKNOWN

- Create a table for reagents used, observations seen, and prediction of which **sets** your unknowns contain from your results in your laboratory notebook.
- List ALL reactions performed and your results in the table in your laboratory notebook.

NOTE:

- You should perform reactions with ALL chemicals suggested.
- Your unknown sample should contain three sets (one cation from each set).

PART II. IDENTIFY THE IONS IN YOUR UNKNOWN

- Create a table for reagents used/experiments performed, observations seen, and prediction of which IONS your unknown contains from your results in your laboratory notebook.
- List ALL reactions/experiments and their results in the table in your laboratory notebook.
- Make sure you record ALL observations of reactions performed for this experiment.
- From the three sets determined in Part I and the observations/results from Part II, utilize your results to determine the three ions present in your unknown sample.

Considerations for Discussion and Conclusions Section

As you write your discussion and conclusions paragraphs in your lab notebook, here are some tips and questions you should be answering. (This is in addition to the "Guidelines for a good discussion and conclusion" handout provided to you at your first lab class.)

- Explain the process by which you were able to identify the cations present in your unknown sample (sets and cations).
  - Were you able to eliminate the presence of one set based on your results for Part I?
  - Were you able to eliminate the presence of some ions based on your results for Part II?
- Identify which three cations you believe to be present in your unknown sample.
- Utilize the results you obtained to verify the presence of these cations. You should use more than one experimental result for this purpose. Note: A result such as "No Precipitate" can also assist to verify the presence (or absence) of a set or cation for Parts I and II.

G.C.2-4

A Freezing Point Depression Study

You will need a watch/timer capable of measuring seconds for this experiment.

This is an adaptation of an experiment developed by the Woodrow Wilson National Fellowship Foundation, Princeton, New Jersey.

Objectives

By the end of this experiment you should be able to

- calculate molal concentrations
- calculate the freezing point depression of a solution
- calculate the molecular weight of a substance from its freezing point depression

References

✓ Read Chapter 13.4 in your textbook.

Pre-Lab Instruction

- Review the technique regarding the handling of a freezing point apparatus.

Background

Colligative properties are properties of solutions that depend on the amount of solute particles in the solution (concentration) and are independent of the nature of the solute. Freezing point depression, boiling point elevation, vapor pressure lowering, and osmotic pressure are all colligative properties, as explained in Chapter 13.4 of the textbook.

If a solute is dissolved in a solvent, when the solution is cooled, it does not begin to freeze until the temperature has gone beyond the freezing temperature of the pure solvent.

During today's experiment, we will determine the freezing points of water as a pure solvent and three different substances sodium chloride (electrolyte), sucrose (nonelectrolyte), and ethylene glycol (nonelectrolyte). For the sodium chloride and the sucrose studies, we will calculate the van't Hoff factor, "i," and for ethylene glycol, we will calculate the molecular weight/molar mass from the collected freezing point data.

Since colligative properties depend on the amount of solute particles in the solution, the amount of freezing point depression can be calculated using the following equation:

$$\Delta T_f = k_f \times m \times i \qquad \text{Eq. 1}$$

where

ΔT_f is the change in the freezing point ($T_{solution} - T_{pure\ solvent}$ = negative number)

k_f is the molal freezing-point-depression constant, which depends on the solvent

m is the molal concentration of the solute particles
i is the van't Hoff factor, which, for dilute solutions, is the number of dissolved ions per mole of solute (for example, i = 5 for $Al_2(SO_4)_3$), and is explained on page 496 in the textbook.

The concentration of the solution (m) is expressed in *molality*, which is the number of moles of solute per kilogram of solvent. Thus, the units for molality are **mol/kg**. Freezing point depression can also be used to determine the molecular weight (MW) or molar mass of the solute. This is explained in detail in your textbook, Chapter 13 on pp. 471–473, including examples. You may have to refer to the textbook for an example of calculations to determine the molar mass/molecular weight of the solute.

Chemicals and Special Equipment

Ice, table salt, ethylene glycol, sodium chloride (NaCl) as Kosher Salt, and sucrose ($C_{12}H_{22}O_{11}$)
Thermometer, large test tubes with wire stirrer, stopper, and LabQuest 2

HAZARD

Ethylene glycol is highly toxic. Acetone is both flammable and volatile. Please use near a fume hood.

Procedure

You will work with a lab partner. Each of you will carry out three freezing point determinations: water, either an NaCl or a sucrose solution, and ethylene glycol. You will combine and compare your results after you are done with your data collection. Please make sure you collect the results from your lab partner before you leave the lab, as you are required to perform ALL calculations associated with this lab experiment.

Preparation of an Ice Bath

Fill up the 600 mL beaker three-fourth full with ice and approximately 100 mL of water. Cover the surface of the ice with **table salt**. Stir this ice–salt mixture with a stirring rod and make sure the temperature drops to between −5.0°C and −7.0°C. If the temperature does not drop, do not panic; adjust the amount of water and salt in the bath.

Data Collection

Prepare your Vernier LabQuest 2 interface for data collection. Note: there is an On/Off button.

1. Press and release the power button located on the top left-hand edge of the LabQuest 2 interface device. You should see "the screen initializing, please wait." This may take a few minutes. When the file sensor screen appears, this indicates a successful power up.
2. Connect the stainless steel temperature probe to channel 1 of LabQuest 2 interface device. LabQuest App will auto-ID the connected sensor and automatically set the default collection rate for the sensor.
3. Ch1: Temperature will appear with the current temperature reading.
4. Click on "Rate" on the screen and change to 1 sample/second. Click "Done."
5. Click on the DURATION icon on the screen and change it to 500 seconds and click "Done."
6. You are now ready to collect data. Click on the File icon and go to "New" (discard old data if it appears on the screen).
7. Click on the **green** triangle (**Δ**) on the left hand bottom of the screen to begin data collection.
8. Data collection begins and LabQuest graphs the data in real time. It may take several seconds for the probe to equilibrate to the temperature of the solution.
9. After the designated time, press on the RED box in the lower left hand to stop the data collection.

10. To get the temperature min/max to display on the screen: tap analyze Statistics → **Temperature**
11. Rinse and dry the probe.
12. To turn off the LabQuest 2, go to the home icon and press to access the system screen. Go to the shut down icon and press it then okay and it will shut down.

Determination of Freezing Points of Solutions

Check out a freezing point apparatus from your TA. The freezing point apparatus is shown in Figure 4-1. Depending on your assignment, perform the steps either for the NaCl, the Kosher Salt, or for the sucrose. Weigh out about 5.80 g of NaCl (Kosher Salt) and write the exact measurement in your lab report. In your 250 mL beaker, mix 100 mL of deionized (DI) water and the 5.80 g of NaCl (Kosher Salt). Mix until all crystals dissolve. Set it aside for future use. Weigh out about 34.0 g of sucrose and write the exact measurement in your lab report. In your 250 mL beaker, mix 100 mL of DI water and 34.0 g of sucrose. Mix until all crystals dissolve. Set it aside for future use.

Procedure for Determining the Freezing Point of a Solution

To test the freezing point of the pure solvent, fill the freezing point apparatus at least half way with DI water and place it in the ice bath. Stir the water gently with the stirrer while keeping track of the temperature. Record the temperature every 30 seconds. When the **first ice crystals** appear on the inside wall of the test tube, record the temperature. Do not wait until the entire sample is frozen. This is the freezing point of the pure solvent. Discard the water and fill up the test tube half way with your next solution, either NaCl or sucrose.

Repeat the procedure above with the prepared NaCl or sucrose solutions. These solutions can be poured down the drain when you have finished.

Molecular Mass Determination from Freezing Point Depression

Measure **6.20 g** of ethylene glycol using a 10 mL graduated cylinder on the scale and mix it with 100 mL of DI water in your 250 mL beaker. Fill up the large test tube at least half way with your water/ethylene glycol mixture and repeat the procedure above for determining the freezing point of this solution. Make sure to record the freezing point temperature **when the first ice crystals appear**. Do not wait until the entire sample is frozen.

Once the experiment is complete, dispose of the ethylene glycol in a properly labeled waste container in the fume hood.

Figure 4-1. Freezing point apparatus

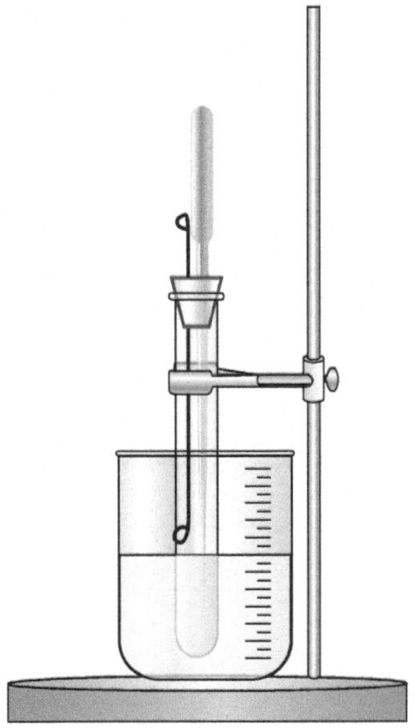

G. C. 2-4: LABORATORY REPORT

Name: ____

Lab partner's names: ______________________________________

Units, calculations, and significant figures will be counted toward your grade!
Freezing point of DI water

Time, min.	Temperature, °C
0.5	
1.0	
1.5	
2.0	
2.5	
3.0	
3.5	
4.0	
4.5	
5.0	
5.5	
6.0	
6.5	
7.0	
7.5	
8.0	
8.5	
9.0	
9.5	
10.0	
10.5	
11.0	
11.5	
12.0	

Time, min.	Temperature, °C
0.5	
1.0	
1.5	
2.0	
2.5	
3.0	
3.5	
4.0	
4.5	
5.0	
5.5	
6.0	
6.5	
7.0	
7.5	
8.0	
8.5	
9.0	
9.5	
10.0	
10.5	
11.0	
11.5	
12.0	

Time, min.	Temperature, °C
0.5	
1.0	
1.5	
2.0	
2.5	
3.0	
3.5	
4.0	
4.5	
5.0	
5.5	
6.0	
6.5	
7.0	
7.5	
8.0	
8.5	
9.0	
9.5	
10.0	
10.5	
11.0	
11.5	
12.0	

Freezing point of _______
Freezing point of ethylene glycol

Part I: Determination of van't Hoff factor "i" for Sodium Chloride and Sucrose

Please perform calculations in your laboratory notebook.

Data from your lab partner must be shared and collected during class time and signed off by your TA.
Freezing point of DI water, °C: ____________________

Collected Data	Sodium Chloride	Sucrose
Freezing point (°C)		
ΔT_f		
Mass of solute (g)		
Number of moles of solute		
Mass of solvent (kg)		
Molality (mol/kg)		
Calculation of van't Hoff factor "i"		

Part II: Determination of Molecular Weight for Ethylene Glycol

Freezing Point, °C	ΔT_f	Mass of Solute, g	Mass of Solvent, kg	Calculated Molecular Weight, g/mol

Calculations, Conversions, and Values

1. Calculate the molality of the sodium chloride and the sucrose solutions.
2. For the mass of solvent, assume that the **density of water is 1.0 g/mL.**
3. Determine the value of "i," where **K_f for water = −1.86°C/m (m = molal)**. Hint: Use Equation 1.
4. Calculate the molecular mass of ethylene glycol based on the freezing point depression. You may want to refer to your textbook for this calculation.

G. C. 2-4: LABORATORY REPORT

CONSIDERATIONS FOR DISCUSSION AND CONCLUSIONS SECTION

As you write your discussion and conclusions paragraphs in your lab notebook, here are some tips and questions you should be answering. (This is in addition to the "Guidelines for a good discussion and conclusion" handout provided to you at your first lab class.

- When you prepared the ice bath, you put salt on the ice. Why was salting the ice necessary in order to complete the experiment, and how does this apply to the principles/concepts behind this experiment?
- Use your results and discuss the theory behind the experiment (including colligative properties, electrolytes and nonelectrolytes, etc.).
- When you are discussing sources of error:
  - Determine the theoretical molecular weight of ethylene glycol (from the molecular formula) and compare your experimental data (obtained in the lab) with the theoretical data.
  - In calculating the molecular weight of ethylene glycol, what value did you assume "i" to equal? Explain why you made this assumption.

G.C.2-5

Rate of Reaction Study

You will need a watch with capabilities of timing seconds for this experiment!

Objectives

By the end of this experiment you should be able to

- calculate the rate of a reaction
- calculate the overall order of a reaction
- calculate the order of a reaction with respect to a particular reactant or product
- recognize the effect of changes, such as concentration, temperature, or addition of a catalyst, on the reaction rate

References

✓ Read Chapter 14.1–14.6 in your textbook.

Background Information

The rate of a reaction is a positive quantity that expresses how the concentration of a reactant or product changes with time. On a macroscopic level, rates of reactions are determined experimentally, and several factors may affect the rate of a reaction: reactant concentration, temperature, reaction mechanism, and presence or absence of a catalyst. Three of these factors will be explored in this experiment.

Today, we will demonstrate how changes in reactant concentration, temperature, and catalyst presence can affect the rate of a reaction. In Part I, the effect of varying the temperature on the rate of a reaction is determined (see 14.1 in the textbook). In Part II, the effect of varying concentration on the rate of a reaction will be determined. The same reaction will be run four times with varying concentrations of reactants H_2O_2, $2I^-$, and $2H^+$ in each run, and the time it takes for the reaction to complete will be measured. In Part III, the effect of a catalyst on the rate of a reaction is measured.

The reaction to be studied in today's experiment is

$$\mathbf{H_2O_2 + 2I^- + 2H^+ \rightarrow I_2 + 2H_2O} \qquad \text{Eq. 1}$$

One factor must be accounted for in this type of experiment. All reactions slow down as their reactants are being used up. To get a true picture of the rate of the reaction, it is best to measure only the initial rate of the reaction. Therefore, the reaction conditions are designed to detect how long it takes to use up **some** of the reactants. To do this, starch and thiosulfate ions are added to the reaction solution. Starch binds to molecular iodine (I_2) produced in the above reaction, and produces a deep blue-black color. As such, it can behave as an indicator for this reaction. When the blue color first begins to appear, we know that I_2 has just begun to be formed in the solution. The thiosulfate ions prevent the blue color from appearing immediately. Thiosulfate ions ($S_2O_3^{2-}$) react very quickly with I_2 by the following equation:

$$\mathbf{2S_2O_3^{2-} + I_2 \rightarrow 2I^- + S_4O_6^{2-}} \qquad \text{Eq. 2}$$

Thiosulfate ions react with the I_2 before it has a chance to bind to the starch. So as long as thiosulfate ions remain in solution, the blue color will not appear; at the point when all the thiosulfate ions are consumed, the I_2 being produced will bind to the starch and the blue color will appear. The reaction shown in Equation 2 **is not** part of the main reaction. It is a "side reaction" that does not affect the main reaction. However, knowing the number of moles of thiosulfate ions consumed by the reaction tells us how many moles of I_2 were produced by the main reaction before the blue color appeared. Knowing the time it took for the blue color to appear allows us to **calculate the rate of the reaction** with units: moles of I_2/L × time, min.

The **rate of a reaction depends on reactant concentrations** and can be expressed in an exponential equation. We will use the results obtained in this experiment to calculate the "order" of the reaction. The order of a reaction shows how the reaction is dependent on the concentration of each reactant (review pages 515–518 in the text). For a reaction that contains three reactants—A, B, and C—the rate of reaction is shown in Equation 3.

$$\textbf{Rate of reaction} = \mathbf{k\,[A]^p\,[B]^q\,[C]^z} \qquad \text{Eq. 3}$$

where ^{p} is the order of the reaction with respect to A, and ^{q} is the order of the reaction with respect to B etc. When the appropriate values for k and the exponents (p, q, and z) are put in place, this equation can be called the "rate law" for the reaction. Each exponent tells how much effect changing the concentration of a particular reactant will have on the rate of the reaction. For this reaction, these exponents are expected to be whole integers and they express the *reaction's order with respect to a reactant.* Note: **p, q, and z are not the coefficients** in the reaction equation. This is a rate equation, not an equilibrium equation.

Reaction Rate and Concentration

In this experiment, to determine how concentration changes affect the rate of a reaction, you will be making two different reaction mixtures, differing only in the concentration of one reactant. For this experiment, the concentration of one of the reactants will be doubled, while the others are kept the same, and the rate of the two trials will be measured as a function of concentration. You can then write rate laws for each trial (review 14.3 in the text). Since for the two reaction mixtures you know the quantities of each reactant and you have determined how the concentrations affected the rate, you can then use this information to **determine the order of the reaction.** The overall order of the reaction is given by the sum of all the exponents in the rate law equation for that reaction.

$$\textbf{Order} = \mathbf{p + q + z} \qquad \text{Eq. 4}$$

Chemicals and Special Equipment

0.50 M buffer (acetic acid/sodium acetate), 0.30 M acetic acid, 0.50 M potassium iodide (KI), 1% starch solution, 0.10 M sodium thiosulfate ($Na_2S_2O_3$), 0.80 M hydrogen peroxide (H_2O_2) (3%), 0.10 M Cu^{2+} solution, ice, test tubes, piece of white paper, thermometer

Hazard

H_2O_2 is caustic. This solution may be concentrated enough to cause minor burns to the skin. $Na_2S_2O_3$ is an irritant. Avoid contact. Acetic acid is corrosive.

PROCEDURE

Note: Be very careful not to contaminate your solutions, especially the hydrogen peroxide (H_2O_2) and the starch solutions. **Clean and dry all glassware with acetone before using it.** Make sure to dispose of the acetone rinses properly in a waste container.

You will be working in groups of two for this experiment. Prepare an ice bath in your 600 mL beaker. Label three 125 mL and two 250 mL Erlenmeyer flasks with numbers 1 through 5.

Based on Table 5-1, calculate and record in the table the total amount you will need of each solution. Keep in mind that you will use the composition presented for Flask no. 1 twice. With your lab partner, clean and label beakers (50 mL–400 mL) for all the solutions and get the total amount needed from the stock bottles. If you need additional glassware, you can use a 50 mL Erlenmeyer flask. Record the concentration of each on your lab report.

Rinse a 10 mL graduated pipet with water and then rinse with a small amount of KI solution. Measure the KI solution into each of the numbered flasks as shown in Table 5-1. Then, rinse the pipet again with water, followed by the starch solution. Measure out the starch solution into each flask.

Finally, using a 100 mL graduated cylinder, accurately measure out the deionized water, buffer, and acid in that order. **Make sure to add each reagent in the order they are presented in Table 1. Make sure to rinse out your pipet before measuring out a new solution to avoid contamination.**

Use the graduated pipet again to measure **exactly** 2.5 mL of your $Na_2S_2O_3$ solution into each of the five numbered flasks. You have now created your reaction solution. When the H_2O_2 is placed in the test tubes, this will initiate the reaction.

Use the pipet again to measure the H_2O_2 solution **into six new clean, dry test tubes. Do not mix the H_2O_2 into the numbered flasks until you are ready to measure the rate of the reaction**.

Part I: Temperature Effects

Work with flask no. 3 only. Thoroughly chill the reaction solution in the ice bath before you add the H_2O_2 to the solution. The temperature of the **solution should be kept at 6–10°C during the entire length of this experiment.** Record the room temperature and the temperature of the reaction solution after it has been chilled. Quickly pour the peroxide solution into the flask. At the exact moment the H_2O_2 is added to the solution, begin timing the reaction. After mixing, swirl occasionally to keep the starch from settling. Keep the beaker in the ice bath during the reaction. Measure the amount of time from the moment the H_2O_2 is added to the solution until the **first pale blue** color appears in the solution. You may want to place the flask on a piece of white paper to help detect the color change.

Table 5-1. Measures of the various solutions* used for the test

Flask No.	KI (mL)	Starch (mL)	DI water (mL)	Buffer (mL)	Acid (mL)	$Na_2S_2O_3$ (mL)	H_2O_2 (mL)
1	1.75	2.50	48.25	15.00	0	2.50	5.00
2	1.75	2.50	50.75	15.00	0	2.50	**2.50**
3	3.50	2.50	46.50	15.00	0	2.50	5.00
4	3.50	2.50	46.50	15.00	0	2.50	5.00
5	1.75	2.50	23.25	15.00	25.0	2.50	5.00
1 + Cu^{2+}	1.75	2.50	48.25	15.00	0	2.50	5.00
Total							

*Must be added in the order of appearance in the table.

Part II: Concentration Effects

Start to work with flask no. 1. Place the flasks on a piece of white paper to help detect the color change, and prepare to begin timing the reaction. When you are ready, quickly pour the H_2O_2 solution from the first test tube into flask no. 1 and stir vigorously to mix. The moment the H_2O_2 is added to the solution, begin timing the reaction. After mixing, swirl occasionally to keep the starch from settling. Measure the amount of time from the moment the H_2O_2 is added to the solution until the **first pale blue** color appears in the solution.

Repeat the above procedure with flask nos. 2, 4, and 5 and record the data. You can time the start of the reactions so that you can watch over more than one flask at the time.

Part III: Effect of Copper

Prepare the reaction mixture in flask no. 1 one more time. However, before adding the H_2O_2 to the solution, add two drops of the Cu^{2+} solution to the reaction solution. Watch this one very carefully; it tends to go quickly. Measure the amount of time to the nearest second from the moment the H_2O_2 is added to the solution until the **first pale blue** color appears in the solution.

Waste Disposal: Acetone must be collected in the waste container in the fume hood. All other solutions and products can be flushed down the drain with lots of water.

CALCULATIONS, CONVERSIONS, AND VALUES

1. Calculate the concentration of $Na_2S_2O_3$ in the reaction solutions based on the volume used in each reaction flask and the total volume calculated.
2. Calculate the concentration of I_2 (number of moles of I_2 produced/L) based on the stoichiometric ratio as shown in Equation 2.
3. To calculate the concentrations of H_2O_2 and KI.
   - You can use $C_1V_1 = C_2V_2$ for calculations of standard dilutions.
   - The final volume for every reaction solution is 75.0 mL.
   - KI is the source of I^-, so $[KI] = [I^-]$.
4. Convert the measured time into minutes:
   - 5 min 40 sec = 340 sec or 5.67 min.
5. Calculate the rate of reaction for each run using "number of moles of I_2 produced/L · min" (Molarity/Min) as the unit.
6. Calculate the order for each reaction (here are some hints):
   - Arrange the rate equations so that the ratio of the concentrations will be equal or greater than one.
   - Round the obtained ratios to the nearest integer.
   - Round the calculated "p, q, or z" to the nearest integer.
   - You may want to consult your textbook for an example (see pp. 515–518).
7. Calculate the overall order of the reaction using Equation 4.

G. C. 2-5: LABORATORY REPORT

Name: ____________________

YOU MUST SHOW ALL YOUR CALCULATIONS TO RECEIVE CREDIT FOR THEM!

CONCENTRATION OF STOCK SOLUTIONS: (M, MOL/L)

Buffer	Acetic Acid	H_2O_2	KI	$Na_2S_2O_3$

Calculation of Concentration of	
$S_2O_3^{2-}$ in mixture	
I_2 produced	

Part I and II

Room temperature __________ Temperature of reaction solution __________

flask No.	time, min	$[H^+]$	$[H_2O_2]$	$[I^-]$	Rate
1		1.76×10^{-5} M			
2		1.76×10^{-5} M			
3		1.76×10^{-5} M			
4		1.76×10^{-5} M			
5		3.52×10^{-5} M			

Note: As seen in Table 1, $[H^+]$ was obtained from the buffer solutions (containing acetic acid and sodium acetate) for flasks 1–4 and from both the buffer as well as the addition of acetic acid for flask 5. These values have been provided to you for this experiment. You can learn how to calculate these values when we cover acids, bases, and buffers, and when we perform the G.C. 2-9 experiment.

Order of Reaction with Respect to the Reactants:

Order with Respect to	Order
H_2O_2 (compare flasks no. 1 and 2)	
I^- (compare flasks no. 4 to 1)	
H^+ (compare flasks no. 5 to 1)	
Overall order of the reaction	

Part III

Work with a Cu^{2+} Catalyst	time, min.	rate of reaction Molarity/Min
Flask no. 1 + Cu^{2+}		

Show all your calculations in your laboratory notebook.

Rate of Reaction = Number of moles of I2 produced per Liter per minute.

G. C. 2-5: LABORATORY REPORT

Considerations for Discussion and Conclusions Section

As you write your discussion and conclusions paragraphs in your lab notebook, here are some tips and questions you should be answering. (This is in addition to the "Guidelines for a good discussion and conclusion" handout provided to you at your first lab class.)

- What is the rate law equation for the reaction?
- What does the rate law equation tell you about the reaction in terms of order of each of the reactants?
- Explain your results in terms of the three experimental conditions tested.
- Why did we ask you to measure the room temperature and temperature of reaction solutions?

G.C.2-6

Determination of a "K_c"

OBJECTIVES

By the end of this experiment you should be able to

- understand the concept of dynamic equilibrium
- establish the equation of K_c for a known reaction
- understand and use Beer's law
- use a spectrophotometer
- calculate the slope from a graph

REFERENCES

- ✓ Read Chapter 15 in your textbook.

PRE-LAB INSTRUCTION

- Review the techniques and tutorial regarding the use of a volumetric flask, pipetting, spectroscopy and the Lab Quest 2 spectrovis spectrophotometer.

BACKGROUND

A reaction is at *equilibrium* when the system has reached a level of stability where the concentration of both the reactants and the products are constant. At equilibrium, the rates of both the forward and the reverse reactions are the same. The system appears to be unchanging since both reactions take place with the same speed. This is why we call it a *dynamic equilibrium*. A system at equilibrium can be described by an *equilibrium constant* (K_c) at a given temperature.

In today's experiment, you will study a reaction between Iron(III) (Fe^{3+}) and thiocyanate ions (SCN^-), in an aqueous solution that produces a complex ion, $Fe(SCN)^{2+}$ through a ligand exchange reaction. Equation 1 shows a simplified form of the reaction.

$$\mathbf{Fe^{3+}{}_{(aq)} + SCN^{-}{}_{(aq)} \leftrightarrows Fe(SCN)^{2+}{}_{(aq)}} \qquad \text{Eq. 1}$$

From the balanced equation, an expression for the equilibrium constant can be determined (see Chapter 15.2 for equilibrium constant expression).

To help you understand equilibrium, Table 6-1 shows the formation of the product for a reaction. This is known as an "ICE" table, where the **initial** concentrations, **change** in concentration, and concentration at **equilibrium** are displayed.

Table 6-1

At the Time of	concentration of		
	Fe^{3+}	SCN^-	$Fe(SCN)^{2+}$
Initial	C_1 mol/L	C_2 mol/L	0 mol/L
Change	−X	−X	+X
Equilibrium	$(C_1 - X)$ mol/L	$(C_2 - .X)$ mol/L	X mol/L

Note: There is a negative sign in front of the concentration of the reactants at the time of change. This arises because the reactants are consumed to produce the $Fe(SCN)^{2+}$ product. If the initial concentration of the reactants and the concentration of the product at equilibrium are known, you can calculate the concentration of the reactants at equilibrium. Using these concentrations at equilibrium, you can then calculate K_C once the expression for equilibrium constant is determined.

The product, ferrithiocyanate $Fe(SCN)^{2+}$, has such an intense red color that this reaction is used to detect small quantities of Fe^{3+}. Since the reactants are colorless and the product has an intense color, you can use a detection method called spectrophotometry to determine the concentration of the product by using Beer's law.

$$\mathbf{A = \varepsilon \times b \times c} \qquad \text{Eq. 2}$$

where "A" is absorbance, "ε" (epsilon) is a constant known as the molar absorptivity of the solute, "b" is the path length of light through the sample, and "c" is the concentration of the product. Since our "b" and "ε" are constant, the absorbance is directly proportional to the molar concentration; thus

$$\mathbf{A = k \times c} \qquad \text{Eq. 3}$$

where $k = \varepsilon \times b$.

For Part I of this experiment, the absorbance of a series of solutions with a known concentration of $Fe(SCN)^{2+}$ will be measured at 450 nm. You will draw a Beer's law graph plotting the absorbance vs. the concentration of the solutions, and the results of this will allow you to **calculate "k" (slope).** Using one of the reactants in excess, you can assume that the limiting reactant will completely react to form a known amount of product. This is based on Le Chatelier's principle, which states that the system will adjust to reduce the effect of a change. For this experiment, Fe (III) will be the limiting reagent.

To **determine K_c** in Part II of this experiment, you will prepare a second series of solutions by varying the concentration of the reactants, but keeping the total volume in the flask equal to 7.0 mL. You can use this dilution information to calculate the **initial concentration of reagents** Fe^{3+} and SCN^-. You will then measure the absorbance of these solutions at 450 nm and using absorbance data collected for the second series and the value calculated for "k" from Part I to calculate the **concentration of products at equilibrium.** You can then use Table 1 to calculate the **concentration of the reagents at equilibrium.** Finally you are equipped with all the information required to allow you to determine "K_c."

Chemicals and Special Equipment

0.0050 M iron III nitrate ($Fe(NO_3)_3$) made in 0.10 M nitric acid (HNO_3), 1.0 M and 0.0050 M potassium thiocyanate (KSCN) made in 0.10 M HNO_3 and 0.10 M HNO_3 solutions, cuvettes, Lab Quest 2 Spectrovis spectrophotometer, graduated pipet, 12 mL test tubes, Kimwipes, graduated pipet, 100.0 mL volumetric flask

Hazard

1.0 M KSCN is an irritant. 0.10 M HNO_3 is corrosive. Avoid skin contact.

WASTE

All reagents should be collected together and discarded in the properly labeled waste containers in the fume hood.

Note: Be careful to use the appropriate concentrations when you prepare your solution series for both Parts I and II. They involve two **different** concentrations of both $Fe(NO_3)_3$ and KSCN.

PROCEDURE

Part I: Determination of "k"

You will be working in groups of two for this experiment. The procedure is written as if you were to work by yourself. To perform Part 1, you will be making a diluted $Fe(NO_3)_3$ solution in a volumetric flask. You will use **this** diluted solution of $Fe(NO_3)_3$, along with other reagents, to create the five different solutions for Part I **only**.

Each group should check out two 10 mL graduated pipets, five test tubes, five cuvettes, and one 100 mL volumetric flask from your teaching assistant (TA) and clean them thoroughly. Label the test tubes 1 through 5 and place them in your 600 mL beaker. Label a 400 mL beaker as your waste container; you can collect all waste solutions together.

Preparation of Diluted Iron (III) Nitrate solution for Part I

Using a graduated pipet transfer exactly 3.0 mL of **0.0050 M $Fe(NO_3)_3$** into a 100 mL volumetric flask. Slowly add enough deionized (DI) water to bring the volume of solution to just below the volume mark on the neck of the flask. Add the last few drops with a Beral pipet. The bottom of the meniscus should line up exactly with the volume mark. Close the flask and invert several times to mix the solution.

Transfer the **diluted** $Fe(NO_3)_3$ **solution** from the volumetric flask into a properly labeled 250 mL beaker. Obtain 35 mL of 0.10 M HNO_3 in a 100 mL beaker. Obtain 25 mL of **1.0 M KSCN** in a 50 mL beaker. Using a graduated pipet prepare your standard solutions according to Table 6-2.

Transfer the respective solutions from the beakers by pipetting into the test tubes, one solution at a time. **Remember to clean and rinse the pipet** with the correct reagent to avoid cross-contamination. Cover the test tubes with parafilm and mix the solutions thoroughly by inverting the test tubes a couple of times.

Using a Beral pipet fill up the cuvettes about three-fourth full with each of the solutions. Wipe each cuvette with a Kimwipe to make sure it is clean and dry. **Do not label the cuvettes**.

1) You will now be using the Lab Quest 2 Spectrovis Spectrometers. Your TA has already calibrated the instrument, and the cursor is set for the appropriate wavelength, at 450 nm. Place the cuvette with 0.1 M HNO_3 in the spectrometer.
   a. Cuvette should be capped or parafilmed.

Table 6-2. Standard solutions using diluted $Fe(NO_3)_3$

Test Tube no.	0.10 M HNO_3 (mL)	Diluted $Fe(NO_3)_3$ (mL)	1.0 M KSCN (mL)
1	4.0	1.0	5.0
2	3.0	2.0	5.0
3	2.0	3.0	5.0
4	1.0	4.0	5.0
5	0.0	5.0	5.0

Do **NOT** insert anything, other than the cuvette, in the slot when the light source in on since that can result in a fire or serious burns. As you have seen in the video clip, insert the most dilute sample into the slot. Make sure the triangle on the cuvette is facing the light source. Record the absorbance reading from the screen. Remember, your absorbance readings should be within 0.00–1.0. Remove your first sample and insert the second one. Repeat this process for all of your samples.

When you are finished with all of your data collection, dispose of the solutions in the waste container and wash your cuvettes with DI water. Draw a Beer's law graph of absorbance vs. concentration of solutions no. 1–5 on Excel and determine "k."

Part II: Determination of "K_c"

Clean and dry the test tubes. Number them 6 through 10. Use the 0.10 M HNO_3 from Part 1 and obtain **15 mL of 0.0050 M KSCN** in a 50 mL beaker. Using a graduated pipet prepare your solutions according to Table 6-3.

Cover the test tubes with parafilm and mix the solutions thoroughly by inverting the test tubes a couple of times. Transfer the solutions into the cuvettes as in Part I. Record the absorbance of each solution at 450 nm. Use the same Lab Quest 2 instrument as before if possible to reduce error in changing instruments. Dispose of the solutions in the waste container when you are finished with your data collection. When you have finished the experiment, clean all of your glassware, including rinsing your cuvettes with DI water, and return the checked-out items. Your TA will not sign your data if you have not returned the items yet. TURN OFF the Lab Quest 2 Spectrovis Spectrometers by unplugging them from the Lab Quest 2 Verniers.

Table 6-3. Solutions for determination of K_c using 0.0050 M $Fe(NO_3)_3$ from TA

Test Tube No.	0.10 M HNO_3 (mL)	0.0050 M $Fe(NO_3)_3$ (mL)	0.0050 M KSCN (mL)
6	5.0	1.0	1.0
7	4.5	1.0	1.5
8	4.0	1.0	2.0
9	3.5	1.0	2.5
10	3.0	1.0	3.0

CALCULATIONS, CONVERSIONS, AND VALUES

1. Calculate the concentration of the **diluted** $Fe(NO_3)_3$ solution for Part I. Note: **You must first calculate the concentration of the diluted solution in the volumetric flask**. Then from this dilution, you use portions to make up different 10.0 mL samples. That is, you diluted two times for the first part of this experiment.
2. Calculate "k" using Equation 4 and the graph you completed. The units will then be L/mol.
3. Calculate the concentration of the product, $[Fe(SCN)^{2+}]$, at equilibrium. Note: You can use the absorbance values of the product and the "k" you calculated with Equation 3 again.
4. Calculate the initial concentrations of reactants Fe^{3+} and SCN^- for solutions 6 − 10. Note: You can use Table 3 to determine how much the total volume of each solution was used.
5. Calculate the concentrations of reactants at equilibrium using Table 1, the ICE table.

G. C. 2-6: LABORATORY REPORT

Name: ______________________________

Lab partner's name: ______________________________

YOU MUST SHOW ALL YOUR CALCULATIONS TO RECEIVE CREDIT FOR THEM!

Proper use of significant figures and units is part of the grading.

Determination of "K" Known Concentrations

Stock concentration of $Fe(NO_3)_3$ ____________

Concentration of $Fe(NO_3)_3$ in 100 mL volumetric flask ____________

Sol. No.	Concentration of $Fe(NO_3)_3$ in each test tube	Absorbance
1		
2		
3		
4		
5		
Calculation of "k" from the manual graph:		Value of "k":

Determination of "K_c" Unknown Concentrations

Sol. No.	Absorbance of $Fe(SCN)^{2+}$	Concentration of product, $[Fe(SCN)^{2+}]$, at equilibrium	K_c*
6			
7			
8			
9			
10			

*Determine only after completing the table above.

*Determine only after completing the table below.

Sol. No.	Initial Concentrations of Reactants		Concentrations of Reactants at Equilibrium	
	$[Fe^{3+}]$	$[SCN^-]$	$[Fe^{3+}]$	$[SCN^-]$
6				
7				
8				
9				
10				

Average value of K_c: ________________

G. C.2-6: LABORATORY REPORT

CONSIDERATIONS FOR DISCUSSION AND CONCLUSIONS SECTION

As you write your discussion and conclusions paragraphs in your lab notebook, here are some tips and questions you should be answering. (This is in addition to the "Guidelines for a good discussion and conclusion" handout provided to you at your first lab class.)

- Explain in your own words how you were able to determine the concentration of reactants at equilibrium? Hint: Discuss the ICE table and how it was used.
- In utilizing your data to explain this experiment, consider and discuss the process by which all data were collected (including graphs).
- For the constant determined at the end, what does the value mean in regard to greater than or less than one, as explained in the text?

Note: As this is an equilibrium experiment, please make sure to explain all concepts in terms of equilibrium.

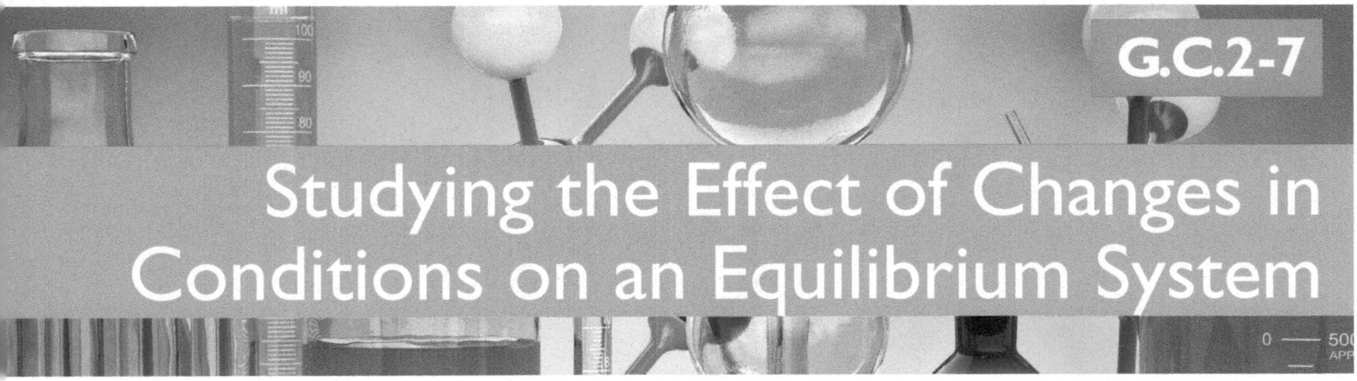

G.C.2-7

Studying the Effect of Changes in Conditions on an Equilibrium System

Objectives

By the end of this experiment you should be able to

- understand the principles of dynamic equilibrium
- explain and predict how changing system conditions affect the equilibrium

References

✓ Read Chapter 15.6 in your textbook.

Background

In today's lab, we will be observing how different equilibrium systems respond to changes in reaction conditions. You will study five different reaction systems and test how these systems react to different types of changes (concentration, temperature, etc.). This relationship was first discovered by Henri Le Chatelier, and is commonly referred to as Le Chatelier's principle (see Chapter 15.6 in the text).

Chemicals and Special Equiptment

0.10 M $HC_2H_3O_2$ solution, solid $NaC_2H_3O_2$, 0.10 M Na_2CrO_4 solution, 1.0 M and 6.0 M concentrated HCl solutions, 1.0 M and 6.0 M NaOH solutions, solid $BiCl_3$ (stored in a desiccator), solid $CoCl_2{\cdot}6H_2O$ (stored in a desiccator), 0.10 M $Fe(NO_3)_3$, 0.10 M KSCN and 0.10 M $AgNO_3$ solutions, universal pH paper

Hazard

0.10 M Na_2CrO_4 solution is slightly toxic. HCl is corrosive. NaOH is caustic. $CoCl_2{\cdot}6H_2O$ crystals can damage your skin. Avoid contact. Notify your teaching assistant (TA) immediately in case of an accident!

Procedure

System No. 1: The Dissociation of Acetic Acid

$$\mathbf{HC_2H_3O_2 \leftrightarrows H^+ + C_2H_3O_2^-}$$

Pour about 2 mL of 0.10 M acetic acid solution into a test tube. To test the pH of the solution, dip a glass stirring rod into the solution and place a drop of it on a piece of universal indicator paper. See Figure 7-1. Record the color of the pH paper and determine the pH according to the chart on the pH paper's container. Then using your spatula, add a small amount of solid $NaC_2H_3O_2$ to the acetic acid and mix it until all the solid dissolves.

Figure 7-1. Measuring pH with universal pH paper

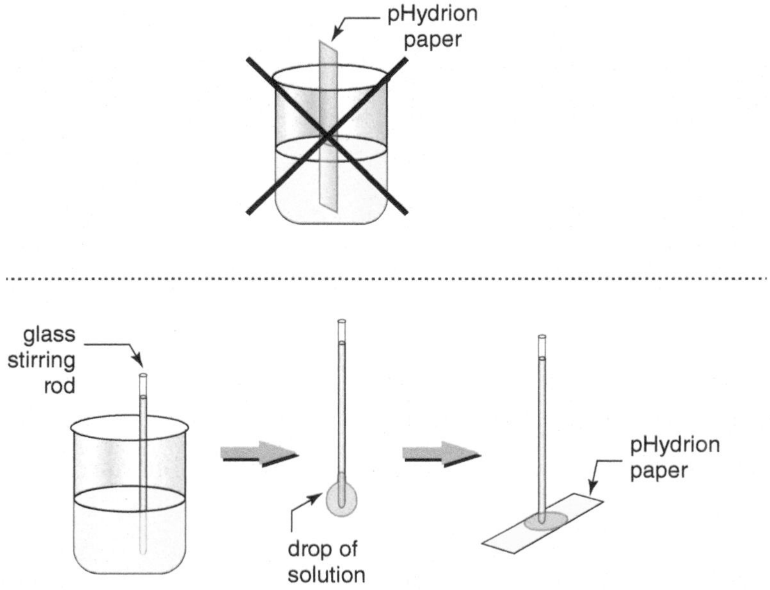

Test the pH of the solution again. Record your observations with reference to the pH. Answer the questions based on your observations with respect to an equilibrium shift.

Waste Collection: Dispose of the solution down the drain.

System No. 2: The Equilibrium of Chromate–Dichromate Ions

$$2CrO_4^{2-} + 2H^+ \rightleftarrows Cr_2O_7^{2-} + H_2O$$

(Note: One of the ions is orange and the other is yellow.)

Pour about 2.5 mL of 0.10 M Na_2CrO_4 solution into a test tube. Record the color of the initial solution. Add one drop of 1.0 M HCl to the test tube and record observation and then add 2 drops of 1.0 M NaOH to the same test tube. Record the colors of the test tubes. If no color change is noticeable, then repeat the above steps until you see some color change at least in one of the two test tubes. Answer the questions based on your observations with respect to an equilibrium shift.

Waste Collection: Dispose of solutions in proper waste container for System 2.

System No. 3: The Equilibrium of $BiCl_3$–BiOCl

$$BiCl_{3(aq)} + H_2O_{(l)} \leftrightarrows 2H^+_{(aq)} + BiOCl_{(s)} + 2Cl^-_{(aq)}$$

(Note: $BiCl_3$ is soluble, i.e., an aqueous solution.)

Pour about 2 mL of water into a 50 mL beaker and add a tiny amount of $BiCl_3$ (on the tip of your spatula) to the water. Add 6.0 M HCl dropwise while continuously stirring until you see any noticeable change. Record your observation. Pour another 25 mL water into the beaker and record your observation.

Waste Collection: Dispose of solutions into proper waste container for System 3.

System No. 4: The Equilibrium of $[Co(H_2O)_6]^{2+}$–$[Co(H_2O)_2Cl_4]^{2-}$ Complex Ions

$$[Co(H_2O)_6]^{2+} + 4Cl^- \leftrightarrows [CoCl_4]^{2-} + 6H_2O$$

(Note: One cobalt complex is pink $[Co(H_2O)_6]^{2+}$ and other is blue $[CoCl_4]^{2-}$. Adding heat will affect the equilibrium, and the reaction will shift to the endothermic direction.)

Prepare a hot water bath. Fill up your 250 mL beaker halfway with water, put 2–3 boiling chips into it, and slowly heat it on the hotplate. Obtain some ice in your 400 mL beaker and prepare an ice water bath by adding water to the ice. Transfer a small amount of $CoCl_2 \cdot 6H_2O$ into a test tube. Add 1–2 mL of water and mix it until the solid is dissolved. Record the color of the solution.

Add concentrated HCl dropwise until a color change occurs. (Note: You should see a color of pure pink or blue. Record your observation.) Add 6.0 M NaOH dropwise until you get a violet (in between light blue and light pink) color. Mix the solution after each drop. Draw about half of the solution into two test tubes. Place one test tube into a hot water bath and the other into the ice water bath for about 10 minutes or until a color change occurs. Record your observations. Keep the water baths to use for the next system.

Waste Collection: Dispose of solutions in proper waste container for System 4.

System No. 5: Equilibrium of the Iron Thiocyanate System

$$Fe^{3+} + SCN^- \leftrightarrows [Fe(SCN)]^{2+}$$

(Note: AgSCN and $Fe(OH)_3$ are insoluble, and the $[Fe(SCN)]^{2+}$ complex ion is blood red, as you have seen in experiment G. C. 2–5.)

Mix about 1.5 mL of 0.1 M $Fe(NO_3)_3$ and 1.5 mL of 0.1 M KSCN solutions in a 50 mL beaker. Add about 20 mL water until the solution becomes lighter red. Draw about 1.5–2.0 mL of the solution into two test tubes. Place one test tube into a hot water bath and the other into the ice water bath for about 10 minutes or until a color change occurs. Record your observations. Transfer about 2.5 mL of the original solution into four test tubes. To the first test tube add 5–10 drops of 0.1 M $Fe(NO_3)_3$. To the second test tube add 5–10 drops of 0.1 M KSCN. To the third test tube add 5–10 drops of 0.1 M $AgNO_3$. To the fourth test tube add 5–10 drops of 6 M NaOH. Record your observations.

Waste Collection: Dispose of solutions in proper waste container for System 5.

G. C. 2-7: LABORATORY REPORT

Name: ______________________________

Answer the following questions in your laboratory notebook in complete detail using Le chatelier's principle to explain and support your observations. Include the direction of the reaction shift in all answers.

System No. 1: $HC_2H_3O_{2(aq)} \leftrightarrows H^+_{(aq)} + C_2H_3O_2{}^-_{(aq)}$

Solutions	pH	Observation
0.1 M $HC_2H_3O_2$		
0.1 M $HC_2H_3O_2$ + $NaC_2H_3O_2$		

1. Did the equilibrium shift and in which direction?
2. How did the pH of the solution change after adding $NaC_2H_3O_2$ to it? (Hint: Identify which ion is being added to the solution and how it affects the concentration of other ions present in solution.)

System No. 2: $2CrO_4{}^{2-}{}_{(aq)} + 2H^+_{(aq)} \leftrightarrows Cr_2O_7{}^{2-}{}_{(aq)} + H_2O_{(l)}$

Solutions	Observations
Na_2CrO_4 solution	
Na_2CrO_4 solution + HCl	
Na_2CrO_4 solution + NaOH	

3. Explain how the equilibrium shifted when you added HCl to the solution?
4. Which chromium ion is orange and which is yellow? How did you know?
5. How did the equilibrium shift when you added NaOH to the solution? (Hint: Think about what NaOH will react with and how that result will affect the other components present in the solution.)

System No. 3: $BiCl_{3(aq)} + H_2O_{(l)} \leftrightarrows 2H^+_{(aq)} + BiOCl_{(s)} + 2Cl^-_{(aq)}$

Solutions	Observations
$BiCl_3$ mixed with water	
Mixture plus HCl	
Mixture diluted with water	

6. Explain how the equilibrium shifted when you dissolved the $BiCl_3$ in excess water.
7. Explain how the equilibrium shifted after the addition of HCl.
8. Explain how the equilibrium shifted after dilution.

System No. 4: $[Co(H_2O)_6]^{2+}{}_{(aq)} + 4Cl^-{}_{(aq)} \leftrightarrows [CoCl_4]^{2-}{}_{(aq)} + 6H_2O_{(l)}$

Solutions	Observations
Dissolved $CoCl_2{\cdot}6H_2O$	
Solution plus HCl	
Heated solution	
Cooled solution	

9. According to the equilibrium equation, which ion(s) do you have after the solid was diluted? Explain how you know the identity of the ion(s) present.
10. Explain what happened when HCl was added to the solution.
11. Explain what happened when the solution was heated. Was the shift in equilibrium exothermic or endothermic?
12. Explain what happened when the solution was cooled. Was the shift in equilibrium exothermic or endothermic?

System No. 5: $Fe^{3+} + SCN^- \leftrightarrows [Fe(SCN)]^{2+}$ **(Forward reaction is exothermic)**

Solutions	Observations
$Fe(NO_3)_3$ + KSCN mixture	
Diluted solution	
Heated solution	
Cooled solution	
Solution + $Fe(NO_{3)3}$	
Solution + KSCN	
Solution + $AgNO_3$	
Solution + NaOH	

13. When the test tube was placed in the hot water bath, was the shift in equilibrium endothermic or exothermic? Explain.
14. Explain what happened when you added more $Fe(NO_3)_3$ and KSCN.
15. Explain what happened after you added $AgNO_3$. Hint: Determine the potential products being formed from adding $AgNO_3$ to the solution.
16. Explain what happened after you added NaOH. Hint: Determine the potential products being formed from adding NaOH to the solution.

G. C. 2-7: LABORATORY REPORT

Considerations for Discussion and Conclusions Section

As you write your discussion and conclusions paragraphs in your lab notebook, here are some tips and questions you should be answering. (This is in addition to the "Guidelines for a good discussion and conclusion" handout provided to you at your first lab class.)
Explain the conditions tested in this lab, including:

- Please use one concentration and one temperature system in your discussion and conclusion section.
- Use examples of conditions that were modified (concentration, temperature, etc.)
- Conclusions you made based on recorded data/observations
- Please look at the experiment as a whole, rather than as a summary of each system.

G.C.2-8

Acid–Base Titrations Using Virtual Lab Software

Objectives

By the end of this experiment you should be able to

- relate equilibrium principles to acids and bases in aqueous solution
- carry out a titration using a pH meter
- draw a titration curve using the obtained data to distinguish between strong and weak acids and bases
- determine the acid dissociation constant K_a of a weak acid

References

✓ Read Chapters 16.1–16.8 and 17.3 in your textbook.

Background

Acids and bases can be categorized as strong or weak based on their ease of ionization of hydronium (H^+) and hydroxide (OH^-) ions. Strength of acids can be measured by determining the pH of the solution: the lower the pH, the greater the $[H_3O^+]$ and hence the stronger the acid. We can quantitatively determine the strength of the acid using the acid equilibrium constant, K_a. Each acid has its own K_a value; the larger the K_a value, the stronger the acid.

In today's experiment, acid–base titrations will be carried out by monitoring changes in the system using a pH meter. You will work in groups of three, using the Virtual Lab software to compare the reaction of three different acids with sodium hydroxide (a strong base), to determine the changes in the pH (and hence $[H_3O^+]$) of the system. The three acids to be titrated include a strong monoprotic acid (hydrochloric acid), a weak polyprotic acid (phosphoric acid), and a weak monoprotic acid (acetic acid).

You will draw an acid–base titration graph, otherwise known as the "titration curve," for each of the above reactions. From the titration curves, one equivalence point for each dissociating H^+ ion can be distinguished. For titration of the diprotic acid, two equivalence points should be observed, one for the dissociation of the first $[H_3O^+]$ and the second for the dissociation of the second $[H_3O^+]$. The pH of the equivalence point in an acid–base titration is taken as the midpoint in the vertical portion of the pH versus volume of titrant curve (Kotz, p. 649). You will be asked to calculate from the pH the $[H_3O^+]$ at different times during the titration. Examples of titration curves and calculations can be seen in Chapter 17.3 of your textbook.

For the titration of the weak acid with a strong base, you will also determine the pK_a from the titration curve, since at the halfway point (half-equivalence point) of the titration, pH is equal to the pK_a. Here, you assume that the starting point is the first equivalence point. Once you have identified the pK_a, you can calculate the K_a of the weak acid.

Chemicals and Special Equipment

Virtual Lab© software developed by Carnegie Mellon University

Procedure

You will work individually, and then share data with two lab partners.

Titration of an acid

Your teaching assistant (TA) will introduce the Virtual Laboratory software and demonstrate the use of it. Choose the appropriate glassware from the drop down menu of glassware icon on the left sidebar on the workbench to carry out the titrations. You will titrate a 10.0 mL sample of each of the acid solutions against a 1.0 M NaOH solution. Prepare the first acid sample and fill up a buret with the NaOH solution. Record the pH value of the acid from the pH meter at the lower right corner of the screen before the titration and after each addition of the base. Add 1.0 mL increments of the NaOH until you see a jump in the pH.

Then start a new titration with a fresh sample of acid, and add the total volume of NaOH solution before the jump occurred at once, then start to add small increments of the base such as 0.10–0.01 mL (e.g., if you notice a jump after adding 7.0 mL, then add 6.0 mL at once, and from then on add 0.1 mL at a time). Continue the titration until the pH of the solution reaches pH $\geq$ 11 and you have at least three data points with the same pH.

Share your data with your lab partners and collect information about the other two acids. Draw a titration curve using Microsoft Excel for the strong acid and diprotic acid, **indicating the pK_a and equivalence points on the graph.** Draw a titration curve manually for the acetic acid. Indicate the pK_a, equivalence point, and halfway point on your graph.

G. C. 2-8: LABORATORY REPORT

Name ____

Lab partners' names: ____________________________________

YOU MUST SHOW ALL YOUR CALCULATIONS TO RECEIVE CREDIT FOR THEM!

Titration of an acid

Concentration of NaOH: ___________

Name of acid: __________

Concentration: _________

Volume of NaOH	pH

Name of acid: ___________

Concentration: __________

Volume of NaOH	pH

Name of acid: ___________

Concentration: ___________

Volume of NaOH	pH

Continuation of Data Collection

Name of acid: ___________

Concentration: __________

Volume of NaOH	pH

Name of acid: ___________

Concentration: ___________

Volume of NaOH	pH

Name of acid: ___________

Concentration: ___________

Volume of NaOH	pH

Continuation of Data Collection

Name of acid: ___________

Concentration: ___________

Volume of NaOH	pH

Name of acid: ___________

Concentration: ___________

Volume of NaOH	pH

G. C. 2-8: LABORATORY REPORT

DATA TABLE AND CALCULATIONS

	HCl	H_2SO_3	CH_3COOH
pH before titration			
pH at equivalence point			
$[H_3O^+]$ before titration			
$[H_3O^+]$ at the equivalence point			

1. What is the pK_a value of the weak acid? _______________
2. What is the K_a value of the weak acid? _______________

Calculations, Conversions, and Values

1. Calculate the hydronium ion concentration, $[H_3O^+]$, from the pH before the titration **and** at the equivalence point for all three acids. Show calculations and record answers on the table above.

2. Indicate the equivalence point(s) on each graph.

3. Indicate the halfway point on the acetic acid graph.

4. Determine the pK_a of acetic acid based on your graph.

5. Calculate K_a of the acetic acid.

G. C. 2-8: Laboratory Report

Considerations for Discussion and Conclusions Section

As you write your discussion and conclusions paragraphs in your lab notebook, here are some tips and questions you should be answering. (This is in addition to the "Guidelines for a good discussion and conclusion" handout provided to you at your first lab class.)

- What is the difference between the three titrations (including differences in equivalence points)?
- From the data collected (K_a, pK_a, etc.), what do these values mean?
- What data were you able to obtain from the graphs drawn and how did they assist in resolving the purpose of the experiment?

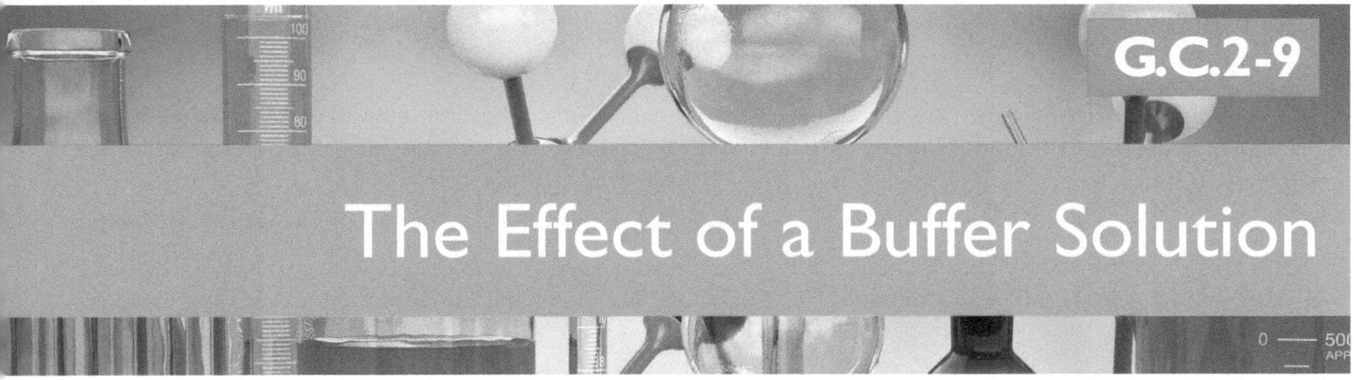

G.C.2-9

The Effect of a Buffer Solution

Objectives

By the end of this experiment you should be able to

- evaluate the pH effect of adding acids and bases to a buffer solution compared to water
- apply the Henderson–Hasselbalch equation to determine the expected pH of a buffer solution
- understand the equilibrium of a buffer system

References

✓ Read Chapter 17.1–17.2 in your textbook.

Background

Buffering systems resist change of pH. The best-known buffer is human blood. Like many delicate systems, the human body is extremely sensitive to pH changes. The hydronium ion concentration, $[H_3O^+]$, in the blood can be affected easily if it is not protected by buffer systems. The normal pH range of human blood is 7.35–7.45. If the buffer does not function and the pH rises above 7.45, the result is a condition called alkalosis. If the pH falls below 7.35, the result is acidosis and can lead to coma and even to death.

Buffer systems are discussed more in detail in Chapter 17.2 in your textbook. A buffer consists of either a solution of a weak acid and the salt of its conjugate base, or a solution of a weak base and the salt of its conjugate acid. For example, we can look at a buffer system made from acetic acid (CH_3COOH) and its sodium salt, sodium acetate ($Na^+CH_3COO^-$). The dissociation of acetic acid is shown below:

$$\mathbf{CH_3COOH_{(aq)} + H_2O_{(l)} \leftrightarrows H_3O^+{}_{(aq)} + CH_3COO^-{}_{(aq)}} \qquad \text{Eq. 1}$$

or

$$\mathbf{HAc \leftrightarrows H^+ + Ac^-} \qquad \text{Eq. 2}$$

where HAc is acetic acid and Ac^- is acetate ion.

The system above will reach equilibrium as indicated by the double arrows in Equation 1. From the equation we can determine the equilibrium constant expression, which, in the case of an acid/base reaction, is referred to as the acid equilibrium constant (K_a) or base equilibrium constant expression (K_b). In the case above, it will be the acid equilibrium constant expression as seen in Equation 3.

The equilibrium constant for acetic acid is expressed as

$$\mathbf{K_a = \frac{[H^+] \times [CH_3COO^-]}{[CH_3COOH]} = 1.8 \times 10^{-5}} \quad \text{or} \quad \mathbf{[H^+] = \frac{K_a \times [HAc]}{[Ac^-]}} \qquad \text{Eq. 3}$$

Understanding what happens in a buffer system is crucial to understanding how to calculate a solution's pH at any time during a titration. In order to understand this concept, think of each species in terms of how it dissociates:

- The weak acid, CH_3COOH, will dissociate to **some** extent (not much, remember that extent is dependent upon the acid's K_a) into CH_3COO^- and H^+.
- The sodium acetate will dissociate into CH_3COO^- and Na^+ ions, but the important species to consider is the CH_3COO^-.

We can calculate the pH of a buffer system by utilizing the **Henderson–Hasselbach equation**:

$$\mathbf{pH = pKa + \log \frac{[acetate]}{[acetic\ acid]}} \qquad \text{Eq. 4}$$

Note: The weaker the acid, the smaller the value of K_a

Additions of Acids and Bases

What happens when a strong acid such as HCl or a strong base such as NaOH is added to the buffer? Each mole of HCl added will be free to react with each **available** mole of CH_3COO^- in solution. We will therefore get the following reaction:

$$\mathbf{H^+ \text{ (from HCl)} + CH_3COO^- \rightarrow CH_3COOH + H_2O} \qquad \text{Eq. 5}$$

The H^+ from HCl reacts with the acetate ions forming additional acetic acid. We get a decrease in acetate ions (since they reacted with the strong acid) and an increase in acetic acid molecules being formed. Note: This reaction can only happen as **long as there are acetate ions still available** in the solution.

Now let us assume that we added 10.0 mL of 1.0 M HCl to the original buffer solution. Since HCl is a strong acid, it dissociates completely. Thus, we can calculate the number of moles of H^+ knowing that there is one mole of H^+ in one mole of dissociated HCl. $1.0\ M \times 0.010\ L = 0.010$ moles of HCl = 0.010 moles of H^+.

A similar thing happens if we add base to this buffer solution. The NaOH would react with the acetic acid; thus its amount would decrease, while more acetate ions would form.

$$\mathbf{OH^- \text{ (from NaOH)} + CH_3COOH \rightarrow CH_3COO^- + H_2O} \qquad \text{Eq. 6}$$

Here are some sample calculations and guidelines to help you along during this experiment. Follow these guidelines until buffer capacity has been reached.

Method:

1. Identify the reaction that occurs between the HCl or NaOH added and the appropriate species in the buffer solution.
2. Assume this reaction goes to completion.
3. Calculate the **amount of moles** of each species before any equilibrium and after each new addition.
4. Calculate the **concentration** of each species before any equilibrium and after each new addition.
5. Apply the equilibrium calculations to determine the pH or use the Henderson–Hasselbalch equation.

Example: 0.50 L of a buffer solution consisting of a 0.1 M solution of acetic acid and a 0.15 M solution of sodium acetate. Using the Henderson–Hasselbach equation, we can determine the pH of the original buffer. $pH = pKa + \log (0.15\ M/0.10\ M) = -\log (1.8 \times 10^{-5}) + \log (0.15\ M/0.10\ M)$. **The pH of the original buffer is 4.92.**

To this buffer solution we are adding 0.01 L of 1.0 M HCl. We can utilize the five steps above to solve.

Step 1: Identify the first reaction:

$$H^+ \text{ (from HCl)} + CH_3COO^- \rightarrow CH_3COOH + H_2O$$

Step 2: Assume it goes to completion (Note: The following is NOT an ICE table.)

	H^+ (from HCl)	**+ CH_3COO^- →**	**CH_3COOH**	**+ H_2O**
Initial number of moles	(0.01 L) (1.0 M) = 0.01 moles	(0.500 L)(0.15 M) = 0.075 moles	(0.500 L)(0.1 M) = 0.05 moles	
Change	−0.01 moles	−0.01 moles	+0.01 moles	
Step 3: Final number of moles	0 moles	0.065 moles	0.060 moles	
Step 4: Final concentration	0 M	(0.065 mol/0.510 L) = 0.127 M	(0.060 mol/0.510 L) = 0.118 M	

Step 5: Apply the equilibrium calculations (see next step, or use the Henderson–Hasselbalch equation). **For the equilibrium calculations**, use the equilibrium equation of the weak acid dissociation and work an ICE table:

	CH_3COOH	**+ H_2O**	**H^+**	**+ CH_3COO-**
Initial	0.118 M		0	0.127 M
Change	−x		+x	+x
Equilibrium	0.118 − x		X	0.127 + x

If we apply the equilibrium calculations utilizing the modified version of Equation 3, we are able to solve for K_a. Note: **The concentration values included are those at equilibrium.**

$$K_a = \frac{x\,(0.127 + x)}{(0.118 - x)} = 1.74 \times 10^{-5}$$

Utilizing the equation above, we can solve for x as the concentration of H^+ or $[H^+]$, and then solve for pH. An alternative approach would be to apply the Henderson–Hasselbalch equation.

$$\text{pH} = \text{pKa} + \log \frac{[\text{acetate}]}{[\text{acetic acid}]} = 4.74 + \log (0.127/0.118) = 4.77$$

As you can see, the change in the pH is minor from 4.92 to 4.77. If you compare the addition of the same amount of HCl to water, which is essentially a dilution of the strong acid into water, the pH would change from 7 to 1.7.

Similar calculations can be performed for a base.
Step 1: Identify the reaction: OH^- (from NaOH) + CH_3COOH à CH_3COO^- + H_2O
Step 2: Assume it goes to completion

	OH^- (from NaOH)	+ CH_3COOH →	CH_3COO-	+ H_2O
Initial number of moles	(0.01 L) (1.0 M) = 0.01 moles	0.500 L)(0.10 M) = 0.050 moles	(0.500 L)(0.15 M) = 0.075 moles	
Change	−0.01 moles	−0.01 moles	+0.01 moles	
Step 3: Final number of moles	0 moles	0.04 moles	0.085 moles	
Step 4: Final concentration	0 M	(0.040 mol/0.510 L) = 0.078 M	(0.085 mol/0.510 L) = 0.167 M	

Step 5: Apply the equilibrium calculations or apply the Henderson–Hasselbalch equation. We apply the equation

$$pH = pKa + \log \frac{[\text{acetate}]}{[\text{acetic acid}]} = 4.74 + \log (0.167/0.078) = 5.07$$

The pH of the new solution is 5.07. As we have shown, a buffer is capable of maintaining a steady pH range of a system even after adding a substantial amount of acid or base. However, it cannot handle large amounts of addition. At some point a buffer will not be able to counteract the increased acid or base concentration. This is when the buffer reaches its "buffer capacity" (see pp. 640–642 of textbook).

In determining the pH of a buffer solution and the pH after addition of a strong acid or base, it is important to understand that the Henderson–Hasselbalch equation will **only** be used in the context of a weak acid dissociation. Once buffer capacity has been reached, the Henderson–Hasselbalch equation will no longer be used.

At Buffer Capacity

At buffer capacity, the moles of Ac^- are equal to the moles of H^+, resulting in **all** the Ac^- ions being consumed. We can also utilize the five steps above to solve; however, for Step 5, the equilibrium calculations will have to be used, since the Henderson–Hasselbalch equation can only be used for buffer systems. For this step, we can substitute $[H^+]$ for $[Ac^-]$, and then solve for $[H^+]$ to determine the pH at buffer capacity as seen in Equation 7.

$$\mathbf{K_a = \frac{[H^+] \times [H^+]}{[HAc]} \quad or \quad [H^+]^2 = Ka \times [HAc]} \qquad \text{Eq. 7}$$

The buffer capacity can also be reached by addition of a base. The base will react with all of the acetic acid, and the acetic acid concentration will be zero. Here the solution contains only acetate ions—a weak base (anions that are conjugate bases of weak acids act themselves as weak bases in aqueous solutions), where the [HAc] will be equal to $[OH^-]$. The pH can be determined using the equilibrium expression for a weak base in Equation 9. We have to keep in mind that we cannot use K_a, because it is the equilibrium constant of an acid system. The use of K_b becomes necessary, and Equation 10 indicates the relationship we need to utilize.

$$\mathbf{K_w = K_a \times K_b} \qquad \text{Eq. 9}$$

$$\mathbf{K_b = \frac{[OH] \times [OH]}{[Ac^-]} \quad or \quad [OH^-]^2 = K_b \times [Ac]} \qquad \text{Eq. 8}$$

After Buffer Capacity

Once the system exceeds the buffer capacity, addition of extra acid or base will change the pH significantly, since there are no more acetate ions to buffer the strength of the strong acid. The system will act as a strong acid or base system, where the acid or base has been diluted. The acid concentration will be equal to $[H^+]$, or in case of a base addition, the base concentration will be equal to $[OH^-]$ once the dilution is taken into consideration. Thus, the pH of the system can be calculated using the simple pH equation:

$$\mathbf{pH = -\log [H^+]} \quad \text{Eq. 10}$$

In today's experiment, you will use a series of indicators and compare the effects of adding HCl and NaOH solutions on the pH of distilled water and on the pH of a buffer using the Virtual Lab software. The disadvantage of using a single indicator is that we can only determine an estimated range of the pH and not an accurate measurement. For more precise determinations, a pH meter is used.

For an easier data comparison, you will draw a graph of pH versus the volume of HCl or NaOH added, for both your distilled water and buffer solutions. An example of this graph is shown in Figure 9-1.

Chemicals and Special Equipment

Virtual Lab© software developed by Carnegie Mellon University

Procedure

Part I. Effects of HCl additions

The link for the virtual lab:

http://www.chemcollective.org/chem/fau/gc29n.html

Choose the appropriate glassware from the drop down menu of the glassware icon on the left sidebar on the workbench. You will need one 100mL volumetric flask and two 250mL Erlenmeyer flasks, one for the buffer and one for the DI water. You will also need two burets, one for the HCl and one for the NaOH.

Prepare 100.0 mL of the buffer by mixing 10.00mL of 1.0 M NaC2H3O2 (Ac-) with 10.00 mL 1.0 M HC2H3O2 (HAc) and then fill up the flask with DI water to the 100mL mark. Using a graduated cylinder, transfer **50.00mL** of the buffer into one of the Erlenmeyer flasks, and 50.00mL DI water into the other one. Name the flasks appropriately to avoid confusion, by right clicking on them and selecting the rename option.

Figure 9-1. pH versus volume of added acid or base to distilled water (¾) and to a buffer solution (Å)

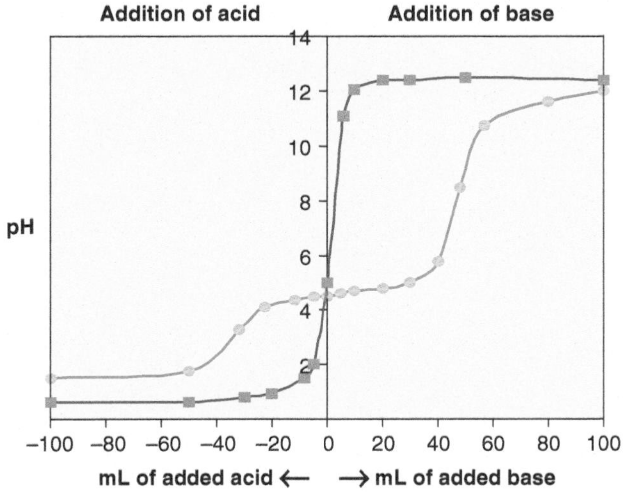

(Note: volume transferred for dilution calculations). There no indicators will be used in this lab, please do not observe color change.

Fill up a buret with 0.1 M HCl solution. From the pH meter at the lower left corner of the screen record the pH value of the buffer and of the water before the first and after each addition of the acid.

Using the buret, transfer 5.00 mL of 0.1 M HCl (there is no 0.1 M solution in the stockroom, you have to dilute 1.0 M HCl to 0.1 M HCl using M1.V1=M2.V2 formula) into the buffer flask and record the pH of the solution. Continue by adding the following amounts of 0.1M HCl to the buffer solution as seen in your data table: 5.00 mL, 10.00 mL, 10.00 mL, 20.00 mL, 50.00 mL, while recording the pH and color of the solution.

Using the buret, transfer 5.00 mL of 0.1M HCl into the water flask by transferring 5.00 mL of 0.1M HCl into the water flask and record the pH of the solution. Continue by adding the following amounts of 0.1M HCl to the water solution as seen in your data table: 5.00 mL, 10.00 mL, 10.00 mL, 20.00 mL, 50.00 mL added into the water flask, while recording the pH of the solution.

Part II. Effects of NaOH Additions

Using a graduated cylinder, transfer 50.00mL of the buffer into one of the Erlenmeyer flasks, and **50.00mL** DI water into the another flask. Again, name the flasks appropriately to avoid confusion, by right clicking on them and selecting the rename option. Fill up a buret with 0.10M NaOH solution (there is no 0.1 M solution in the stockroom, you have to dilute 1.0 M HCl to 0.1 M HCl using M1.V1=M2.V2). From the pH meter at the lower left corner of the screen record the pH value of the buffer and of the water before the first and after each addition of the base.

Using the buret, transfer 5.00 mL of 0.1 M NaOH into the buffer flask and record the pH. Continue by adding the following amounts of 0.1M NaOH to the buffer solution as seen in your data table: 5.00 mL, 10.00 mL, 10.00 mL, 20.00 mL, 50.00 mL, while recording the pH of the solution.

Using the buret, transfer 5.00 mL of 0.1M NaOH into the water flask by transferring 5.00 mL of 0.1M NaOH into the water flask and record the pH of the solution. Continue by adding the following amounts of 0.1M NaOH to the water solution as seen in your data table: 5.00 mL, 10.00 mL, 10.00 mL, 20.00 mL, 50.00 mL added into the water flask, while recording the pH of the solution. Dispose of your used buffer, water and sodium hydroxide solutions.

CALCULATIONS, CONVERSIONS, AND VALUES

1. Calculate the experimentally determined $[H^+]$ (M) from the pH data obtained by using the pH meter.
2. Calculate the expected pH (what are you expecting to obtain based on the concentrations of the components in each solution) for each addition, for both the buffer solution and the water when adding acid or base. Use the equations in the reading.
3. Draw a "pH vs. the volume of added **acid** to distilled water and buffer" graph.
4. Draw a "pH vs. the volume of added **base** to distilled water and buffer" graph.

G. C. 2-9: LABORATORY REPORT

Name: ____

YOU MUST SHOW ALL YOUR CALCULATIONS TO RECEIVE CREDIT FOR THEM!

Part I. Addition of 0.100 M HCl to Buffer and Deionized Water

Concentration of buffer: For acetic acid: ____________ For acetate: ________________

Volume of HCl	Total Volume of HCl	pH Meter Reading		Expected pH (Calculated)	
		Buffer	Water	Buffer	Water
0	0				
5.00	5.00				
5.00					
10.00					
10.00					
20.00					
50.00					

Calculations must be shown in your laboratory notebook.

Part II. Addition of 0.100 M NaOH to Buffer and Deionized water

Volume of NaOH	Total Volume NaOH	pH Meter Reading		Expected pH (Calculated)	
		Buffer	Water	Buffer	Water
0	0				
5.00	5.00				
5.00					
10.55					
10.00					
20.00					
50.00					

Calculations must be shown in your laboratory notebook.

G. C. 2-9: LABORATORY REPORT

CONSIDERATIONS FOR DISCUSSION AND CONCLUSIONS SECTION

As you write your discussion and conclusions paragraphs in your lab notebook, here are some tips and questions you should be answering. (This is in addition to the "Guidelines for a good discussion and conclusion" handout provided to you at your first lab class.)

- Briefly explain/discuss all the concepts involved in the experiment (Hint: There are at least three).
- Discuss the methods used to solve for the pH.
- Compare the pH response of adding HCl and NaOH to the two systems tested using the graphs drawn.

G.C.2-10

Determination of K_{sp}

OBJECTIVES

By the end of this experiment you should be able to

- establish an equation to express the solubility constant of a salt
- titrate a saturated $Ca(OH)_2$ solution with a standard HCl solution
- determine the $[OH^-]$ for the saturated $Ca(OH)_2$ solution
- calculate the K_{sp} of $Ca(OH)_2$
- relate the solubility of a compound based on the K_{sp} value

REFERENCES

✓ Read Chapter 17.4–17.7 in your textbook.

PRE-LAB INSTRUCTION

- Review the technique regarding the handling of a buret and a pipet.

BACKGROUND INFORMATION

In General Chemistry I Laboratory you already learned about solubility rules, and earlier this semester, we discussed that the solubility of a certain compound changes in relation to certain factors, such as temperature. Compounds precipitate when one of the products is insoluble in water. Even though the compound precipitates, a very small amount of the compound will dissolve in water. An equilibrium is established where the ions of the insoluble salt continuously go in and out of solution. Quantitatively, the amount of that insoluble compound which can be dissolved in aqueous solution at a given temperature can be determined by K_{sp}, the solubility product constant. All compounds, even the highly soluble sodium chloride, have a K_{sp}. However, the K_{sp} of a compound is commonly considered only in cases where the compound is very slightly soluble and the amount of dissolved ions is not simple to measure.

Your primary objective in today's experiment is to test a saturated solution of $Ca(OH)_2$ and use your observations and measurements to calculate the K_{sp} of the compound. You will do this by titrating the prepared $Ca(OH)_2$ solution with a standard HCl solution. By determining the molar concentration of dissolved hydroxide ions in the saturated $Ca(OH)_2$ solution, you will have the necessary information to calculate the K_{sp}.

The dissolution reaction is shown below.

$$Ca(OH)_2 \longleftrightarrow Ca^{2+}{}_{(aq)} + 2OH^-{}_{(aq)}$$

The equilibrium expression for $Ca(OH)_2$ is

$$K_{sp} = [Ca^{2+}]\,[OH^-]^2$$

Chemicals and Special Equipment

LabQuest 2 computer interface, Vernier pH sensor, ring stand and ring, two 100 mL beakers, two 250 mL beakers, two 50 mL graduated cylinders, 50 mL buret, saturated calcium hydroxide ($Ca(OH)_2$) solution, 0.050 M hydrochloric acid (HCl) solution, (optional) indicator solution, buret clamp, filter paper, filter funnel, magnetic stirrer and stirring bar, and computer

Hazard

HCl is corrosive and toxic. $Ca(OH)_2$ is caustic. Avoid spilling these on hands or clothing.

Procedure

1. Obtain and wear goggles.
2. Obtain about 30 mL of a saturated $Ca(OH)_2$ solution.
3. Measure out about 15 mL of the filtered solution into a 100 mL beaker. Record the precise volume of $Ca(OH)_2$ solution that you are using.
4. Obtain about 50 mL of 0.050 M HCl solution.
5. Connect a pH sensor to Channel 1 of the Vernier LabQuest 2 computer interface.
6. Set up the beaker of $Ca(OH)_2$ solution on a magnetic stirrer. If you are not using a magnetic stirrer, use a glass stirring rod to stir the solution throughout the titration.
7. Use a utility clamp to suspend the pH sensor on a ring stand, as shown in Figure 10-1. Position the pH sensor in the $Ca(OH)_2$ solution and adjust its position so that the sensor is not struck by the magnetic stirring bar.
8. Connect a buret to the ring stand. Rinse and fill the buret with the 0.050 M HCl solution.
9. Start the Logger Pro program on your computer. Open the file "23 K_{sp}" from the *Advanced Chemistry with Vernier* folder.
10. Conduct the titration carefully. The guidelines below are suggestions; use your judgment in conducting the titrations to get the best results.

Figure 10-1

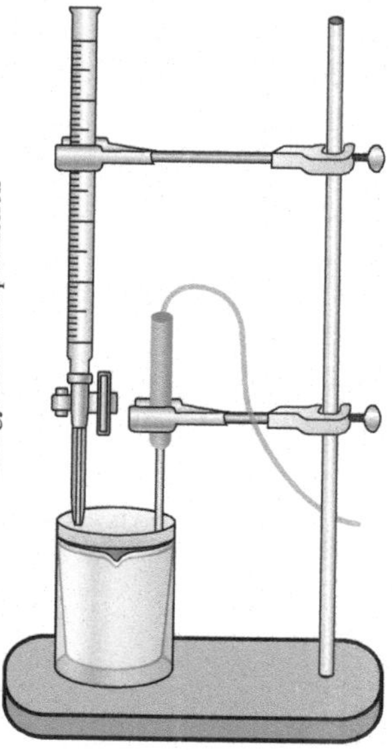

a) Before you have added any of the HCl titrant, click [collect] and monitor pH for 5–10 seconds. Once the displayed pH reading has stabilized, click [keep]. In the edit box, type 0 (for 0.00 mL added). Click [OK] to continue.

b) Add a small amount of the titrant, up to 0.50 mL. When the pH stabilizes click [keep]. Enter the current buret reading. Click [OK] to continue.

c) Continue adding the HCl solution in increments that lower the pH consistently, and enter the buret reading after each increment.

d) When you reach the equivalence point, continue adding the HCl solution until the pH value remains constant.

11. When you have finished collecting data, click [stop]. Choose Store Latest Run from the Experiment menu to save the results of the first trial.
12. Follow the steps below to find the *equivalence point*, which is the largest increase in pH upon the addition of a very small amount of HCl solution. A good method of determining the precise equivalence point is to take the second derivative of the pH-volume data, a plot of $\Delta^2pH/\Delta vol^2$.

e) View a plot of the second derivative by clicking on the Next Page button.

f) Analyze the second derivative plot and record the volume of HCl at the equivalence point.

13. Dispose of the reaction mixture as directed. Rinse the pH sensor with distilled water in preparation for the second titration.
14. Repeat the necessary steps to titrate a second sample of the filtered $Ca(OH)_2$ solution.
15. Rince the pH probe with water and return it to it's container.
16. To turn off the Lab Quest 2, go to the Home icon and press to access the system screen. Go to the shut down icon and press it then go to okay and it will shut down.

Waste Collection

1. To dissolve the unused precipitate in the test tubes, obtain 20 mL of 10% HCl solution and transfer into the two test tubes containing the solid. Discharge as much of the HCl solution and precipitate as possible into your waste beaker.
2. Rinse the test tubes with two portions of deionized water from your water beaker and dispose of the rinses into your waste beaker.
3. Transfer all waste generated from this lab into the properly labeled waste container.

G. C. 2-10: LABORATORY REPORT

Name ____

Lab partners' names: ____

YOU MUST SHOW ALL CALCULATIONS TO RECEIVE CREDIT FOR THEM!

Titration of a base

Concentration of HCl: ___________

Name of base: ___________

Volume of HCl	pH

Name of base: ___________

Volume of HCl	pH

DATA TABLE: TITRATION OF THE SUPERNATANT LIQUID

Data	Trial 1	Trial 2
Volume of $CaOH_2$ solution, mL (supernatant liquid)		
Initial buret reading, mL		
Final buret reading, mL		
Volume of HCl required for titration, mL		
Equivalence point (mL)		
Average K_{sp}		

G. C. 2-10: LABORATORY REPORT

Name: ______________________ **Temperature of water:** ________________

YOU MUST SHOW ALL YOUR CALCULATIONS TO RECEIVE CREDIT FOR THEM!

DATA ANALYSIS AND CONCLUSIONS FOR DISCUSSION

1. Calculate the $[OH^-]$ from the results of your titrations. Explain your calculations.

2. Calculate the $[Ca^{2+}]$. Explain your calculations.

3. Calculate the K_{sp} for $Ca(OH)_2$. Explain your calculations.

4. Temperature affects equilibrium in one direction or the other (depending on whether the system is exo or endothermic). Discuss.

5. What does the value of K_{sp} tell you in terms of equilibrium?

6. The theoretical value for K_{sp} of $Ca(OH)_2$ at 25°C is 9.0×10^{-6}. Compare and suggest reasons for the difference.